SpringerBriefs in Food, Health, and Nutrition

More information about this series at http://www.springer.com/series/10203

Hong-Sik Hwang

Advances in NMR Spectroscopy for Lipid Oxidation Assessment

Springer

Hong-Sik Hwang
Agricultural Research Service
United States Department of Agriculture
Peoria, IL, USA

ISSN 2197-571X ISSN 2197-5728 (electronic)
SpringerBriefs in Food, Health, and Nutrition
ISBN 978-3-319-54195-2 ISBN 978-3-319-54196-9 (eBook)
DOI 10.1007/978-3-319-54196-9

Library of Congress Control Number: 2017933956

Printed on acid-free paper

This Springer imprint is published by Springer Nature
The registered company is Springer International Publishing AG
The registered company address is: Gewerbestrasse 11, 6330 Cham, Switzerland

Abstract

Although there are many analytical methods developed for the assessment of lipid oxidation, different analytical methods often give different, sometimes even contradictory, results. The reason for this inconsistency is that although there are many different kinds of oxidation products, most methods measure only one kind of oxidation product. For the quality control of food products and for better understanding of the factors affecting lipid oxidation, it is necessary to improve the current methods and to develop new analytical methods that provide more accurate assessment of lipid oxidation. NMR spectroscopy techniques including ^{1}H, ^{13}C, and ^{31}P NMR are very powerful and reliable tools to determine the level of lipid oxidation, to identify oxidation products, and to elucidate oxidation mechanism. ^{1}H NMR spectroscopy has demonstrated its reliability, accuracy, convenience, and advantages over conventional analytical methods in the determination of the level of oxidation of edible oils during frying and storage by monitoring changes in several proton signals of oil including olefinic, bisallylic and allylic protons. This modern analytical method has been used to identify oxidation products, including primary oxidation products such as hydroperoxides and conjugated dienes and secondary products such as aldehydes, ketones, epoxides, alcohols, dimers and polymers, and their derivatives. By identifying intermediates and final oxidation products, mechanisms for lipid oxidation were elucidated. Another type of NMR method, ^{13}C NMR, has also been used to identify oxidation products. The relatively newer method, ^{31}P NMR spectroscopy, can also provide additional information on the molecular structure of an oxidation product. Backgrounds, principles, advantages over conventional methods, most recent advances, and future prospects of these methods will be discussed.

Keywords Lipid oxidation • NMR • Oxidation products • Edible oil • Vegetable oil • Oxidation mechanism

Acknowledgements

I thank Matthew I. Hwang and Esther Y. Hwang for improving the use of English in and organization of the text.

Contents

Chapter 1
Conventional Analytical Methods to Assess Lipid Oxidation

Lipid oxidation is a major cause of the deterioration of quality in food, and accurate qualitative and quantitative assessment of lipid oxidation is important for quality assurance of food products (Shahidi and Zhong 2010; Pignitter and Somoza 2012). Lipid oxidation occurs during manufacturing, transportation, storage, cooking, and frying of edible oils and oil-containing foods. Although sensory analysis is the most reliable method to evaluate the extent of lipid oxidation, it is not practical for routine analyses and generally lacks reproducibility (Gray 1978). For this reason, numerous chemical and physical analytical methods have been developed to assess lipid oxidation, including conjugated diene value, per oxide value, alcohols, epoxides, *p*-anisidine assay, HBR titration, iodometric titration, xylenol orange, total polar compounds (TPC), high performance liquid chromatography (HPLC), fatty acid composition determined by gas chromatography-mass spectrometry (GC-MS), Fourier transform infrared spectroscopy (FT-IR), volatile products using gas chromatography (e.g. solid phase microextraction), and dimers/polymers by size exclusion chromatography (SEC) (Schaich 2013b). Again, it is inevitable to have such a huge number of analytical methods for the accurate assessment of the level of oxidation. This is because lipid oxidation is a very complicated process involving numerous oxidation products. Compounds formed during lipid oxidation may vary with different oils (different fatty acid compositions), antioxidants (inherent and/or added), oxidation temperatures, contents of water, acid and other minor ingredients since these factors can alter the mechanisms and routes of the oxidation reactions. Unfortunately, despite tremendous efforts on understanding mechanisms of lipid oxidation, lipid oxidation is not completely understood, and it is almost impossible to accurately predict oxidation products under different oxidation conditions.

Among numerous oxidation products, there are two important questions: which oxidation product is the best one to represent the oxidation level of lipid? And which analytical method should be used? The answer depends on the goal of the analysis. Some people may be interested in volatile aldehydes that affect the smell of food

© The Author(s) 2017

H.-S. Hwang, *Advances in NMR Spectroscopy for Lipid Oxidation Assessment*,
SpringerBriefs in Food, Health, and Nutrition, DOI 10.1007/978-3-319-54196-9_1

while some may be interested in non-volatile oxidation products, which remain in food and affect taste and health. It should also be noted that, in practical uses, convenience and quickness of analysis are very important factors in addition to accuracy and reliability. Since most standardized assays typically measure one kind of oxidation product satisfying one of these goals, one should take precaution when selecting standardized assays. Inconsistency in results from different assays is a serious issue of the current analytical methods for lipid oxidation, and there is no ideal method that correlates well with changes in organoleptic properties of oxidized lipids throughout the entire course of oxidation. For this reason, although it is troublesome, it is very common to measure two to four different indications of oxidation at the same time to have more reliable data on the deterioration of oils. To overcome the inconvenience, it has been urged to develop new analytical methods that can combine the concomitant detection of many different oxidation products for a more consistent assessment of lipid oxidation (Pignitter and Somoza 2012).

1.1   Mechanisms of Lipid Oxidation

In the 1940s, there were intensive efforts to understand mechanisms of lipid oxidation and a free-radical mechanism, and the chain reactions of radicals were proposed (Bolland and Koch 1945; Gunstone and Hilditch 1945; Holman and Elmer 1947). The current understanding of lipid oxidation is based on the proposed radical reactions involving initiation, propagation, and termination as shown in Fig. 1.1 (Frankel 2012b). Unsaturated lipids such as oleic, linoleic, and linolenic acids are the most susceptible species to oxidation in food. During lipid oxidation, an unsaturated lipid (LH) loses a hydrogen to form a lipid free radical (L·), which rapidly reacts with oxygen to form peroxyl radicals (LOO·). Peroxyl radicals react with other lipid molecules to produce more lipid radicals in the

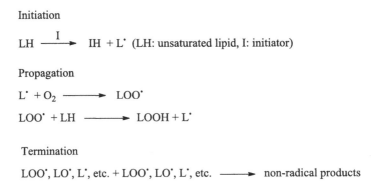

Fig. 1.1 Free radical reaction mechanism of lipid oxidation (Frankel 2012b)

propagation step. Hydroperoxides (LOOH) undergo a variety of reactions, including thermal or metal-catalyzed homolysis, to produce peroxyl (LOO·), alkoxyl radicals (LO·), alkyl (L·), and other radicals. When these highly reactive radicals accumulate, they react with each other to form many kinds of non-radical products in the termination step.

While being widely accepted as the mechanism of lipid oxidation, this free radical chain reaction mechanism cannot explain all the oxidation products. There are on-going efforts to further understand lipid oxidation reactions and courses of reactions other than the hydrogen abstraction in propagation, including β-scission of oxygen, internal rearrangement to epodioxides, addition, disproportionation of lipid peroxyl radicals and many other reactions to alter mechanisms and oxidation products were also proposed (Schaich 2005, 2012, 2013a).

1.2   Methods to Determine Primary Oxidation Products and Their Problems

As shown in Fig. 1.1, hydroperoxides (LOOH) are the major primary reaction products of fatty acids. Hydroperoxides of lipid, which are generally referred to as peroxides, are the compounds responsible for further reactions to produce secondary oxidation products (Gray 1978). It is known that the peroxides themselves do not contribute to the off-aromas causing rancidity but that secondary oxidation products, especially carbonyl compounds, do (Reindl and Stan 1982). Although the peroxides are readily decomposed to other secondary oxidation products in the presence of metals or at high temperatures such as frying temperatures, they are relatively stable at room temperature and in the absence of metals and thus the concentration is built up during the oxidation process (Choe and Min 2006). The concentration of peroxides can be determined by several methods and the peroxide value has become one of the most widely used analytical methods for lipid oxidation (Dobarganes and Velasco 2002). Numerous methods were reported to determine the peroxide value, and among them iodometric methods are very widely used. The iodometric method determines the concentration of iodine produced from the reaction between hydroperoxides and hydrogen iodide, which was produced from potassium iodide and acetic acid (Wheeler 1932; IUPAC 1992). The concentration of iodine can be determined by titration with $Na_2S_2O_3$ (Gray 1978). Figure 1.2 shows the iodometric method in which iodine produced by the reaction of peroxides with hydrogen iodide is determined by titration with $Na_2S_2O_3$. The peroxide value is reported in milliequivalents of O_2 per kilogram of sample.

The iodometric method can be followed by some other detection methods such as the end-point potentiometric determination (Kanner and Rosenthal 1992; Hara and Totani 1988) and the colorimetric detection of I_3^- at 290 or 360 nm (Frankel 2012a; Hicks and Gebicki 1979). However, the iodometric method has the intrinsic problem that iodine can react with double bonds of unsaturated fatty acids and that iodine can be formed by the reaction of potassium iodide with oxygen, which is present in the solution to be titrated (Lea 1952).

Fig. 1.2 Peroxide value
measurement by the
iodometric method using
titration with $Na_2S_2O_3$
(Gray 1978)

Reaction of peroxide with HI

$$KI + CH_3COOH \longrightarrow CH_3COO^-K^+ + HI$$

$$\underset{\text{(peroxide)}}{ROOH} + 2HI \longrightarrow \underset{\text{(alcohol)}}{ROH} + H_2O + I_2$$

Titration of I_2

$$\underset{\text{(purple)}}{I_2} + 2Na_2S_2O_3 \longrightarrow Na_2S_4O_6 + \underset{\text{(no color)}}{2NaI}$$

$$\underset{\text{(peroxide)}}{Fe^{2+} + LOOH + H^+} \longrightarrow Fe^{3+} + H_2O + LO\cdot$$

$$LO\cdot + \text{xylenol orange} \longrightarrow LOH + \text{xylenol orange}\cdot$$

$$\text{xylenol orange}\cdot + Fe^{2+} \longrightarrow \text{xylenol orange} + Fe^{3+}$$

$$LO\cdot + Fe^{2+} + H^+ \longrightarrow Fe^{3+} + LOH$$

$$Fe^{3+} + \text{xylenol orange} \longrightarrow \text{blue-violet complex (560 nm)}$$

Fig. 1.3 Proposed mechanism of peroxide value measurement by determination of the Fe^{3+}-xylenol orange complex (Sochor et al. 2012)

Another class of methods uses the formation of iron complexes in which peroxides oxidize ferrous ions (Fe^{2+}) to ferric ions (Fe^{3+}). Two types of methods are widely used. The first one uses oxidation of ferrous ions in an acidic medium followed by colorimetric determination of ferric ions as ferric thiocyanate, a red-violet complex of strong absorption at 500–510 nm (Shantha and Decker 1994). The other method involves the determination of resulting ferric ions by measuring a blue-purple complex at 550–600 nm formed by ferric ions and the dye, xylenol orange (Jiang et al. 1991). Figure 1.3 shows the proposed mechanism for the method that measures the ferric ion concentration determined by the xylenol orange complex (Sochor et al. 2012).

It should be noted that the test results from different methods depend on the experimental conditions and the reagents used (Gray 1978).

Although the peroxide value is one of the most widely used assays for the assessment of lipid oxidation, there are some problems in addition to the aforementioned problems. First, it does not provide information on secondary oxidation products that more significantly affect the oil quality. Another problem of this method is that the concentration of peroxides reaches a peak and then subsequently declines in a short period of time (Farhoosh and Moosavi 2009). This is because peroxides are consumed quickly to produce a variety of secondary oxidation products such as acids, aldehydes, alcohols, ketones, esters, short-chain hydrocarbons, dimers, and polymers. Therefore, the peroxide value is valid only for the initial stage of oil oxidation at relatively lower temperatures (e.g. storage conditions) and is not valid for harsher conditions such as frying conditions. At any case, the value should be used only before it reaches the peroxide value peaks.

Another type of primary oxidation products, conjugated dienes (or hydroperoxides of conjugated dienes), are formed by the rearrangement of double bonds in polyunsaturated fatty acids (PUFAs) (White 1995; Chatgilialoglu and Ferreri 2010) (Fig. 1.4). Conjugated dienes absorb UV light (230–235 nm), and their concentration can be quantitatively determined by UV spectroscopy. The conjugated diene value is expressed in millimoles per liter of sample (mmol/L). This value overcomes one of the major disadvantages of the peroxide value, the short time to reach the peak value. Farhoosh and Moosavi reported that the conjugated diene value of oil continuously increased during 50 h of frying potato chips in vegetable oil while the peroxide value reached the peak at around 12 h (Farhoosh and Moosavi 2009). The conjugated diene value gave a strong correlation with total polar compounds (TPC), one of the most reliable indications of lipid oxidation ($R^2 = 0.862$) It should also be noted that the conjugated diene method doesn't require additional reagents causing a chemical reaction (for example, for the color development), requires smaller samples, and is quicker, more accurate and simpler than the peroxide value method (St. Angelo et al. 1975). However, this method also has some problems. Conjugated dienes are also primary oxidation products, and therefore, the conjugated diene value eventually reaches a peak value and then declines. Furthermore, this value cannot be taken as the measure of the degree of oxidation unless the fatty acid composition is known since the development of UV chromophores is different with different fatty acids (Gray 1978).

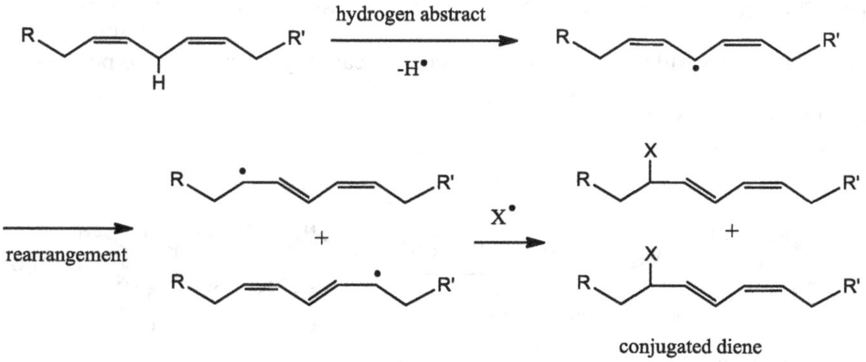

Fig. 1.4 Formation of conjugated dienes

1.3 Methods to Determine Secondary Oxidation Products and Their Problems

The primary oxidation products produce secondary oxidation products through a variety of reactions including cyclization, hydrogen abstraction, addition reaction, recombination, C=C addition, scission, and polymerization (Schaich 2005). Secondary oxidation products include aldehydes, epoxides, ketones, alcohols, carboxylic acids, short-chain hydrocarbons, dimers, oligomers, and polymers (Fig. 1.5). There are numerous analytical methods developed to quantify secondary oxidation products, including colorimetric methods such as TBARS (thiobarbituric acid reactive substances) assay and p-anisidine value and chromatographic methods to measure volatile compounds, fatty acid composition, polymers/oligomers, and total polar compounds (TPC).

The TBARS (thiobarbituric acid reactive substances) assay is one of the widely used classical methods, in which thiobarbituric acid is used as a reagent. Malondialdehyde (MDA) is one of the secondary oxidation products during lipid oxidation and is considered to be a good marker for oxidation. Thiobarbituric acid reacts with malondialdehyde (MDA) and other similar oxidation products as well to produce products absorbing light at 531 nm (Fig. 1.6) (Sinnhuber et al. 1958; Fernández et al. 1997). Since the general procedure usually consists of several steps, including homogenization and centrifugation in an acidic medium and the reaction with thiobarbituric acid at high temperatures (90–100 °C), the possible variation in reaction conditions such as heat treatment exposure time could cause poor repeatability and reproducibility (Barriuso et al. 2013). In addition, the method was designed to detect MDA with thiobarbituric acid. However, it was found that thiobarbituric acid was not very selective to MDA, but it can react with other compounds, such as other aldehydes, amino acids, carbohydrates, and nucleic

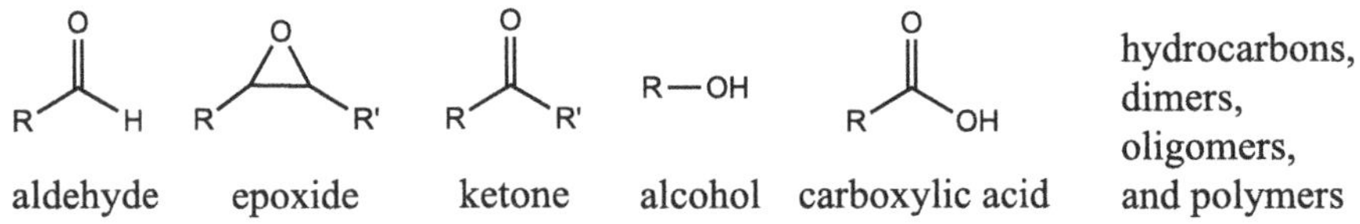

| aldehyde | epoxide | ketone | alcohol | carboxylic acid | hydrocarbons, dimers, oligomers, and polymers |

Fig. 1.5 Secondary oxidation products

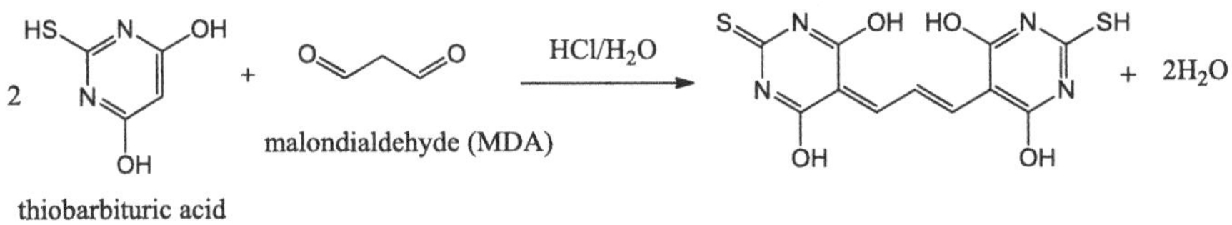

Fig. 1.6 Reaction between thiobarbituric acid and malondialdehyde (Sinnhuber et al. 1958)

acids (Salih et al. 1987; Barriuso et al. 2013). Therefore, the assay is interfered by substances other than oxidation products. Other problems are that MDA can also volatize, and its concentration reaches a peak and then decreases in a short time (Wanasundara et al. 1995).

The p-anisidine value (or anisidine value) is another type of method measuring carbonyl compounds such as aldehydes and ketones produced during the oxidation. The principle of this method is that the amine group in the reagent, p-anisidine, reacts with the carbonyl group of oxidation products to form a Schiff base that absorbs light at 350 nm (Fig. 1.7). In a study using salad oils, a significant correlation was observed between the anisidine value and the flavor score from a taste panel (List et al. 1974). The anisidine value is used to calculate the total oxidation (TOTOX) value along with the peroxide value (TOTOX = 2 × peroxide value + anisidine value) (Sun et al. 2011). The TOTOX value is frequently used to estimate the extent of oxidation in the food industry. The problem of this method is that the colorimetric response of the product from the reaction between the aldehyde and p-anisidine depends on the molecular structure of the product, especially the extent of aldehyde unsaturation. For example, the Schiff base of di-unsaturated aldehyde gives a stronger response than that of a mono-unsaturated aldehyde at the same concentration (Laguerre et al. 2007).

Volatile oxidation compounds are produced by decomposition of hydroperoxides, and the quantities of these compounds can be determined by a chromatographic method, especially by gas chromatography (GC). Volatile aldehydes are the most widely used compounds for the assessment of lipid oxidation, but other volatile compounds such as ketones, alcohols, short carboxylic acids and hydrocarbons can be used as well. Among aldehydes, hexanal is the most frequently used aldehyde, and propanal, pentanal, $(2E, 4E)$-heptadienal, and $(2E, 4E)$-decadienal are also good indicators of lipid oxidation. Since the GC spectroscopy gives signals of many types of oxidation products, it is recommended to use more than one or two markers to express the extent of lipid oxidation (Antolovich et al. 2002). The comprehensive review on the chromatographic methods for volatile compounds was published by Laguerre et al. (2007). There are two typical types of methods to recover volatile compounds for the GC method: extraction and headspace analysis. The simultaneous steam distillation method is widely used as the extraction method, in which the sample is dispersed in water, the extraction solvent is added, and the sample is distilled at an elevated temperature for several hours. Therefore, it is a long and laborious method, and degradation of the volatile compounds during the long process at

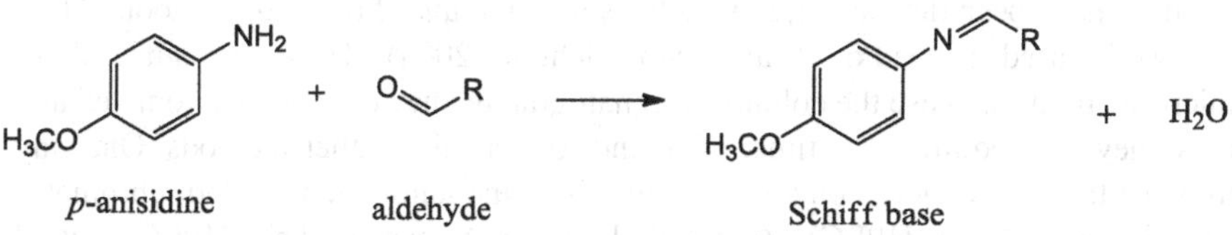

Fig. 1.7   Reaction between p-anisidine and aldehyde

high temperature is a concern. To overcome this problem, the headspace analysis has been developed. Static headspace, dynamic purge-and-trap headspace, and headspace-solid phase microextraction (HS-SPME) techniques are used for this method (Laguerre et al. 2007). The static headspace method uses an airtight vial to collect volatiles, which are in turn injected to the GC column. Since equilibrium is established between the volatile compounds in the headspace and those in the sample (oil), the quantities of volatiles are limited, resulting in a low sensitivity. The quantities of volatiles can be increased with increased incubation time, but it could cause another problem of further degradation of volatiles. The dynamic headspace technique can overcome the problem of the static headspace method since no equilibrium is established in this method. In the dynamic headspace method, the sample vial is continually purged by inert gas to extract volatile compounds, which are, in turn, collected in a porous polymer trap. Since volatiles are constantly released and trapped in the porous polymer trap, the concentration of volatiles is higher than that in the static headspace method. However, some shortcomings of the dynamic headspace method are that the profiles of the volatile compounds can vary with the availability of oxygen in the vial, and that the instrumentation is complex and expensive, and that there are many sources of error (trap drying, trap transfer, purging efficiency, etc.) (Laguerre et al. 2007). The solid phase microextraction (SPME) method is increasingly used since this method takes a shorter amount of time. In this method, headspace volatiles from the sample are adsorbed on a polymeric film coating of a fiber. Then, volatile compounds are released at a high temperature and injected to a GC column. However, one drawback of this method is that the fiber degrades quickly, and at a certain point the fiber must be replaced. Another problem is that the fiber degradation products, such as hexamethylcyclotrisiloxane, octamethlylcyclotetrasiloxane, and decamethylcylcopentasiloxane, are occasionally detected in GC, which could interfere the analysis results (Marsili 2000).

Total polar compounds (TPC) is one of the most reliable analytical methods for lipid oxidation, especially for frying (Gertz 2000; Pignitter and Somoza 2012). In fact, many countries use TPC to set up their regulatory limit for frying oils (typically, 25% TPC) (Rossell 2001). The TPC method uses column chromatography, and the quantity of polar compounds is gravimetrically determined. The procedure for this method involves the following steps: (1) the column is prepared with silica gel, (2) the non-polar fraction is eluted with an eluent, (3) the polar fraction is eluted with another eluent, (4) each fraction is dried, and then (5) the amount of polar compounds is determined. Therefore, the major drawback of TPC analysis is that it requires a lot of time, intensive labor and large amounts of solvents. This problem has been discussed for a long time, and some faster, more economical micromethods have been developed, in which a small amount of oil sample (about 0.08–0.1 g) is used (Sebedio et al. 1986; Schulte 2004). However, since these micro-methods also use the column chromatography, they cannot be automated and thus, they still require more time, labor and solvents than other methods. One way to make this assay more convenient is to utilize the automated modern chromatographic method (e.g. HPLC) (Zhang et al. 2015). A reverse phase HPLC method was introduced by Heain and Isengard (1997), and the results correlated very well with those of the column chromatography. A normal-phase column with a polarity gradient was used to separate polar and non-polar components, in which an evaporative light-scattering detector (ELSD) method was used as a detector (Kaufmann et al. 2001). This research group observed a close correlation between the results measured by using this method and the conventional column chromatography method. An HPTLC (high performance thin layer chromatography) coupled with densitometry was also recently introduced for the determination of polar compounds in heated oil and in frying oil (Correia et al. 2015). In this method, the HPTLC automatically develops the oil sample, and the densitometry allows the quantification of components. This HPTLC-densitometry was fast, accurate, and repeatable for measuring polar compounds. However, for some reason the HPLC methods and the HPTLC-densitometry are not practically used.

Polymers, oligomers, and dimers are continuously formed during lipid oxidation, and the polymerized triacylglycerols are very effective indicators for oxidation. High-performance size exclusion chromatography (HPSEC) (or gel permeation chromatography, GPC) can separate monomers, dimers, oligomers, and polymers. In general, polymerized triacylglycerols (PTAG) comprised of dimers, oligomers, and polymers are measured to determine lipid oxidation, and recently, this method is being used relatively frequently by many research groups (AOAC 1996; Winkler-Moser et al. 2015; Hwang and Winkler-Moser 2014; Summo et al. 2010).

Instead of measuring the appearance of oxidation products, oxidation can be monitored by the disappearance of starting materials such as polyunsaturated fatty acids (PUFAs) (Fhaner et al. 2016). The loss of PUFAs is commonly quantified by means of gas chromatography (GC) after converting triacylglycerols to fatty acid methyl esters (FAME). This method is generally recognized as a reliable method. However, in a study comparing eight different published analytical methods determining fatty acid methyl esters (FAME), it was found that there were significant differences between results obtained by different methods, especially for unsaturated fatty acids (Milinsk et al. 2008). This indicates that this method also needs significant improvement for better consistency.

Some other methods for assessment of lipid oxidation include chemiluminescence (Navas and Jiménez 1996), fluorescence spectroscopy (Karoui and Blecker 2011), infrared (IR) spectroscopy (Kong and Singh 2011), and Raman spectroscopy (Herrero 2008).

Chapter 2
Application of NMR Spectroscopy for Foods and Lipids

Nuclear magnetic resonance (NMR) spectroscopy is one of the most powerful analytical tools to identify organic and bio-organic substances and to elucidate their chemical structures. It is recognized as one of most reliable methods, and it is convenient, fast, and non-destructive. While the proton NMR (^{1}H NMR) is the most frequently used method, it becomes a more powerful tool for analysis of organic compounds when it is used with the carbon NMR (^{13}C NMR) and two dimensional (2D) NMR techniques such as correlation spectroscopy (COSY), nuclear Overhauser effect spectroscopy (NOESY), J-spectroscopy, exchange spectroscopy (EXSY), heteronuclear single quantum coherence spectroscopy (HSQC), and heteronuclear multiple-bond correlation spectroscopy (HMBC).

The NMR spectroscopy has widely been used for analyses of foods such as beer, wine, soy sauce, vinegar, coffee, tea, fruit juice, mandarin oranges, kiwifruit, mangoes, black raspberries, melons, watermelon, tomatoes, lettuce, *Brassica rapa*, potatoes, carrots, maize, wheat, milk, cheese, butter, margarine, honey, fish, and meat (Guillén and Ruiz 2001; Mannina et al. 2012; Marcone et al. 2013; Siddiqui et al. 2017). Compared to chromatographic methods such as GC/MS, LC/UV, and LC/MS, the NMR method has the disadvantage of lower sensitivity. However, the advantages of the NMR method, such as simpler preparation steps, the possibility of obtaining broad information in one measurement, and the high reproducibility, often make it very useful for analyzing food ingredients (Cazor et al. 2006). The NMR method can provide simultaneous access to both qualitative and quantitative information and are being used for purity assessment of organic compounds and identification of potential impurities (Simmler et al. 2014).

Due to such a great versatility of the NMR method, it has been widely used to analyze the quality, structure, composition, characteristics, and ingredients in foods. Moisture in food is often determined by time-domain nuclear magnetic resonance (TD-NMR) or low resolution NMR for the purpose of quality control and quality

© The Author(s) 2017
H.-S. Hwang, *Advances in NMR Spectroscopy for Lipid Oxidation Assessment*,
SpringerBriefs in Food, Health, and Nutrition, DOI 10.1007/978-3-319-54196-9_2

assurance in food industries (Todt et al. 2006). Solid fat contents, the characterization of the ice in frozen food, amounts of biopolymers such as proteins and starch also can be determined using NMR relaxation at low fields (Mariette 2009). The NMR spectroscopy, especially ^{1}H NMR, has been applied to determine the characteristics of oil and fat. For example, Guillén and Ruiz (2003a) determined the amounts of linolenic, linoleic, oleic, and saturated acyl groups in oil using the area of five discrete proton signals of the ^{1}H NMR spectrum. The results were in agreement with the actual amounts of these acyl groups. The NMR method was recognized as a very useful tool for fatty acid composition analysis of oils with additional advantages of shorter time, better convenience, and no chemical modification over traditional analytical methods. Similarly, Castejón et al. (2014) also applied the ^{1}H NMR methodology to the analysis of three different vegetable oils: sunflower, olive, and linseed oils. In this method, it was possible to determine the fatty acid composition in less than 1 min. Furthermore, the accuracy was the same as the traditional method, GC-FID, while the reproducibility was even better. In an attempt to drive forward the automated analysis of the fatty acid composition of edible oils, Castejón et al. (2016) later developed an NMR-based screening method using a 300 MHz NMR instrument, in which the NMR spectra were automatically analyzed and interpreted. This study demonstrated that the complete process from the sample preparation to printing the report takes only 3–4 min.

The amount of free fatty acids in oil can be determined from the areas of the carboxylic group proton (COOH) signal appeared at 11–12 ppm and the methylene proton signal directly adjacent to the carboxyl group resonating and 2.2–2.4 ppm in the ^{1}H NMR spectrum obtained from oil dissolved in a mixture of $CDCl_3$ and DMSO-d_6 (5:1, v/v) (Skiera et al. 2012b, 2014). In the test with a total of 305 oil and fat samples, the NMR method showed a very strong correlation with the conventional method except for hard fat that showed somewhat significant deviations, the data obtained by the two methods were in good agreement. The[1]H NMR method also can be used for. The ^{1}H NMR spectroscopy is also used for analysis of free fatty acids in waxes and oleyloleat, many unsaponifiable materials in oil such as alcohol, sterol, hydrocarbon, and tocopherols. This enables this method to reveal the geographical origin of olive oils and the adulteration of commodity oils (Alonso-Salces et al. 2010). Another important application of the ^{1}H NMR spectroscopy is found in the quality assessment and authenticity of olive oil, which is a frequent target for adulteration and is often diluted with less expensive oils and labeled as pure olive oil (Dais and Hatzakis 2013).

In addition to the common NMR spectroscopy, in which the sample is dissolved in a deuterated solvent, solid-state NMR techniques, which are performed directly on very small sample pieces (typically, a few milligrams) without any chemical or physical manipulation, are also used to examine food structures, chemical compositions, molecular organization, molecular dynamic behavior, and physical properties such as texture and water content (Ribó et al. 2004; Iulianelli and Tavares 2016). For example, the solid-state NMR was used to evaluate the changes in the chemical structure and molecular dynamic, as a response of time storage (Bathista et al. 2012). One of the solid-state NMR techniques, the HRMAS-NMR (high resolution magic angle spinning-nuclear magnetic resonance), was used to assess the metabolic profile of sweet pepper (Ritota et al. 2010). In this study, several compounds, including fatty acids, organic acids, amino acids, and other minor compounds such as trigonelline, C4-substituted pyridine, choline, and cinnamic derivatives in sweet pepper, could be identified.

Chapter 3
¹H NMR Spectroscopy for Assessment of Lipid Oxidation

Due to the aforementioned problems of the conventional analytical methods for the assessment of lipid oxidation, the NMR method has drawn great interest as a new method to assess the level of lipid oxidation. Generally in a review of many methods, the highest reliability is observed when the analytical method measures the following: (1) major products; for example, polymers and total polar compounds (TPC) produced in several percents rather than minor products detected in miliequivalent/kg, mmol/L or mmol, (2) products continuously produced during the course of lipid oxidation; for example, volatile compounds instead of products whose concentration show a peak and decline, or (3) the disappearance of starting materials such as polyunsaturated fatty acids (Hwang 2015). Since the ¹H NMR method typically monitors the disappearance of starting materials by measuring the decrease in signal intensity of specific protons (e.g. olefinic, bisallylic, and allylic protons) of fatty acids during the oxidation process, it was found to be very reliable (Pignitter and Somoza 2012). The reactions of fatty acids continue until all the reactive double bonds disappear. Thus, the NMR method can be used for the entire course of lipid oxidation. Additionally, in the NMR method no chemical reaction with a reagent is involved, but only dissolution of the sample in a deuterated solvent (typically, CDCl₃) is involved, which results in the higher reproducibility and repeatability.

3.1 Assessment of Lipid Oxidation During Oil Storage

It is very important to accurately assess oxidation of edible oils and foods containing lipids during storage, packaging, and transportation. Therefore, studies on the assessment of lipid oxidation are often conducted at storage temperatures, or at

© The Author(s) 2017   15
H.-S. Hwang, *Advances in NMR Spectroscopy for Lipid Oxidation Assessment*,
SpringerBriefs in Food, Health, and Nutrition, DOI 10.1007/978-3-319-54196-9_3

temperatures somewhat higher than typical storage temperatures to accelerate the experiment or to examine it at processing temperatures.

The ¹H NMR spectroscopy technique that assesses oil oxidation was first introduced by Saito (1987), using fish oil heated at 40 °C. Saito found that the standard method, the peroxide value, was not suitable to monitor the oxidation of fish oil for a long period of storage time because the peroxide value reaches the peak value in a short time due to the high content of unstable polyunsaturated fatty acids (PUFA). Saito made an attempt to devise a more reliable analytical method for evaluating the oxidative deterioration of fish oils and their methyl ester during storage and monitored some signals in the ¹H NMR spectrum of fish oil. In this study, it was found that the ratios of olefinic protons (Ro) (4.9–5.8 ppm) to aliphatic protons (0.5–3.0 ppm) decreased markedly as the oxidative deterioration proceeded. Strong correlations were observed between the peroxide value and Ro values (r = −0.972 for pollock liver oil, r = −0.995 for sardine oil, r = −0.955 for the methyl ester of pollock liver oil, and r = −0.988 for the methyl ester of sardine oil). The study also demonstrated the possibility that the peroxide value could be determined from the correlation curve between the peroxide value and the ratio of Ro to the percentage of PUFA in the fresh oil. Oxidation of other fish oils, including saury and mackerel oils stored at 40 °C, was also monitored by the change in the olefinic proton signal (Ro) (Saito and Nakamura 1990) and showed the strong correlations with the peroxide value. Unlike the peroxide value that reached the peak and began to decrease in 7 days, the Ro value indicating the disappearance of PUFA decreased continuously until the end of the storage time of this study (15 days). Therefore, in addition to convenience, the NMR method has an advantage over the peroxide value because it can evaluate the oxidation of fish oil beyond the time limit of the peroxide value.

However, the range of the aliphatic proton signal (0.5–3.0 ppm) contained the diallylmethylene protons (2.6–3.0 ppm) which also markedly decreased during the oxidative deterioration of fish oil. For this reason, Saito and Nakamura (1989) modified the method to use the region of 0.6–2.5 ppm for the aliphatic proton signal and monitored the ratios of olefinic protons (Ro) and diallylmethylene protons (Rm) to the aliphatic protons during storage of pollock oil and sardine oil at 40 °C. Correlation tests with the peroxide value showed strong correlations (r = −0.971 for Ro of pollock oil, r = −0.977 for Rm of pollock oil, r = −0.986 for Ro for sardine oil, and r = −0.974 for Rm of sardine oil). Although the correlation coefficients were found to be similar to those obtained by the previous method in this specific study, the method using the range of 0.6–2.5 ppm for the aliphatic proton signal should generally be more accurate by excluding the region of diallylmethylene protons.

In a separate study, the Saito research group (Saito and Udagawa 1992) utilized the ¹H NMR method to assess oxidation of fish meal stored at 20 °C for 50 days. They compared the NMR method with three traditional analytical methods: peroxide value, acid value, and carbonyl value. The acid value did not change for 50 days, indicating that this value was not suitable for monitoring the oxidation of fish meal. The peroxide value rapidly increased and then began to decrease in 10 days, while the carbonyl value increased over 20 days and then decreased. This result indicated that the carbonyl value was the best method among the three methods, but it could not be used beyond 20 days under the experimental conditions in this study. Compared to these conventional methods, the NMR method showed the important advantage that it could be used for the longer time. Ro (the ratio of olefinic protons to the aliphatic protons) and Rm (the ratio of diallylmethylene protons to the aliphatic protons) continuously decreased during 50-day storage.

Another research group also evaluated the ^{1}H NMR method for assessment of oil oxidation under accelerated storage conditions (Wanasundara and Shahidi 1993), in which canola oil and soybean oil were heated at 65 °C in the dark for 30 days. In this study, the relative number of protons was calculated on the basis of integration of methylene protons of the triacylglycerol backbone at 4.0–4.4 ppm. As shown in Table 3.1, the relative numbers of olefinic protons and diallylmethylene protons of canola oil decreased by 26.8% and 54.6%, respectively. Meanwhile, the number of aliphatic protons increased by 11.2% during 30-day storage at 65 °C. Similarly, soybean oil showed a 27.6% decrease in the number of olefinic protons, a 35.0% decrease of diallylmethylene protons, and a 13.7% increase in diallylmethylene and aliphatic protons. This research group used ratios of aliphatic protons to olefinic protons (Rao) and aliphatic protons to diallylmethylene protons (Rad) to follow the oxidation of these oils, which were simply reciprocal to the values used by Saito and Udagawa (1992). They used slightly different ranges for aliphatic protons (0.6–2.5 ppm), diallylmethylene protons (2.6–2.9 ppm), and olefinic protons (5.1–5.6 ppm) than ranges used by the Saito research group. However, these slight differences would not greatly affect the values. After 30 days of heating at 65 °C, Rao values of canola oil and soybean oil increased from 11.1 to 16.9 and from 9.3 to 14.6, respectively. Due to the smaller number of diallylmethylene protons of canola oil compared to soybean oil, Rad value of canola oil was greater than that of soybean oil and increased from 36.4 to 89.1 while that of soybean oil increased from 18.89 to 33.1 after 30 days heating. Plots of Rao and Rad values against corresponding TOTOX values (TOTOX

Table 3.1 Change of olefinic protons, diallylmethylene protons, and aliphatic protons relative to methylene protons of the triacylglycerol backbone in ^{1}H NMR spectra of canola and soybean oils during storage at 65 °C (Wanasundara and Shahidi 1993; Hwang and Bakota 2015)

	0 day	30 day	Change, %
Canola oil			
Olefinic protons	7.12	5.21	−26.8%
Diallylmethylene protons	2.18	0.99	−54.6%
Aliphatic protons	79.28	88.18	11.2%
Rao	11.1	16.9	
Rad	36.4	89.1	
Soybean oil			
Olefinic protons	8.12	5.88	−27.6%
Diallylmethylene protons	4.00	2.60	−35.0%
Aliphatic protons	75.64	86.00	13.7%
Rao	9.3	14.6	
Rad	18.9	33.1	

= 2 × peroxide value + anisidine value) showed strong correlations (r = 0.984 for Rao and 0.933 for Rad for canola oil and 0.985 for Rao and 0.969 for Rad for soybean oil). This research group concluded that the NMR methodology could be an effective means to simultaneously estimate the overall oxidative changed in vegetable oils.

Shahidi et al. (1994) also used this NMR method with a slight modification, in which the total number of protons of a peak was determined on the basis of integration of terminal methyl protons instead of methylene protons of the triacylglycerol backbone, for the assessment of oxidation of seal blubber oil and cod liver oil stored at 65 °C. The results from this NMR method showed strong correlations with TOTOX value (r = 0.931 and 0.975 for Rao of seal blubber oil and cod liver oil, respectively: r = 0.945 and 0.994 for Rad of seal blubber oil and cod liver oil, respectively). This method was also used for canola and soybean oils heated at 65 °C for 30 days (Wanasundara and Shahidi 1993). In addition, this study showed strong correlations between the NMR values and the TOTOX values. It was also found that while the traditional analytical methods, peroxide value and TBARS (thiobarbituric acid reactive substances) value, reached a peak in 5–15 days, Rao and Rad values increased steadily over the entire length of storage period, 30 days. This NMR method was practically used to measure oxidation of sesame oil heated at 65 °C along with fatty acid composition, iodine value, peroxide value, conjugated diene value, para-ansidine value, and 2-thiobarbituric acid (TBA) value (Abou-Gharbia et al. 1996, 1997). In a study determining Rao and Rad values of borage and evening primrose oils at 60 °C, good correlations with conjugated diene value (r = 0.950–0.995) and TBARS (r = 0.972–0.982) were observed (Senanayake and Shahidi 1999).

Signals other than olefinic proton and diallylmethylene proton signals were also used to monitor the oxidation of oils. Falch et al. (2004) monitored the changes in peaks in the region of 8–10.5 ppm in the ¹H NMR of ethyl ester of docosahexaenoic acid stored at 25 °C in the dark. In the NMR spectrum of the fresh ethyl docosahexaenoate, no peaks were observed in this region. When oxidation started, multiples peaks corresponding to the secondary oxidation products, including aldehyde peaks (CHO) of a variety of aldehydes, appeared at around 9.5 ppm. There were more peaks in this downfield region (8–10.5 ppm), and it was found that the sum of the peak areas in this region could be a good indication of oxidation. Strong correlations between the total peak area and conventional analytic values were observed (R^2 = 0.97 with peroxide value, R^2 = 0.95 with the conjugated diene value, and R^2 = 0.97 with TBARS).

Knothe and Kenar (2004) developed a method to quantitate unsaturated fatty acids in triacylglycerols from the integration values of the allylic and bisallylic (= diallylmethylene) proton signals. This method can be used to determine the oxidative deterioration of oil. For example, Tyl et al. (2008) calculated the content of *n-3* PUFA in fish oils such as eicosapentaenoic acid (EPA) and docosahexaenoic acid (DHA), using the signals in the ¹H NMR spectrum. Then, they oxidized fish oils including capelin oil, cod liver oil, tuna oil, and a blend of fish oils under four different conditions: in the dark at room temperature, with a UV lamp at room temperature, in the dark at 40 °C, and under argon at 40 °C. The concentrations of

EPA and DHA were determined by the NMR method, and it was found that the results were in agreement with the data obtained by gas chromatography (GC), the traditional method. They also found that the ratio of olefinic protons to aliphatic protons (Rao) had a high correlation with the peroxide value ($r = 0.942$), and the signal of the bisallylic (= diallylmethylene) protons showed the same behavior as the olefinic protons.

The method to determine fatty acid composition from the ^{1}H NMR spectrum developed by Knothe and Kenar (2004) was also adapted by Wang et al. (2014) for their study on antioxidant activities of 4-vinylsyringol, α-tocopherol, and sinapic acid in soybean oil at 60 °C. They determined the molar concentrations of linoleic acid (18:2) and linolenic acid (18:3) using the NMR signals, and these values were used to determine the effectiveness of antioxidants during the heating study. The results obtained by the NMR method were in agreement with other traditional methods, peroxide value and TBARS.

Skiera et al. (2012a) improved the methodology of the ^{1}H NMR spectroscopy so that the hydroperoxide (OOH) signal could be used as the marker of oil oxidation. The signal of the hydroperoxide (OOH) appears as a very broad peak when the most common NMR solvent, CDCl$_3$, is used, and it is difficult to accurately integrate the peak. This research group found that the peak was broadened due to hydrogen-bonding between OOH groups. Therefore, when they change the solvent that can interfere with the hydrogen-bonding between OOH groups, the peak became sharper. The sharpest OOH signal was observed with the mixture of CDCl$_3$ and DMSO-d$_6$ (5:1, v:v) among other solvent systems tested. Subsequently, when they measured the concentration of peroxides in different mixtures of a peroxide-free rapeseed oil and the oxidized linoleic acid methyl ester containing a known amount of hydroperoxides (0.91 mg/kg), they found that the NMR method using the intensity of the hydroperoxide signal is a very reliable method. All the measured values were within the 95% prediction bands, indicating the high precision of this NMR method with the mixture of CDCl$_3$ and DMSO-d$_6$ (5:1, v:v) as a solvent. Furthermore, correlation tests showed that the NMR method had a strong correlation with the peroxide value determined by the Wheeler method (Barthel and Grosch 1974). For the practical application, the peroxide values of 290 edible oil samples were analyzed with this NMR method, and the results were compared with the peroxide values from the Wheeler method. The two methods showed strong correlations for sunflower, rapeseed, nut, corn, and thistle oils. However, the data from the ^{1}H NMR method of black seed oil, pumpkin seed oil and olive oil showed significant deviations from the peroxide value measured by the Wheeler method. It was found that the iodometric titration in the Wheeler method was significantly influenced by the essential oil contained in black seed oil (about 0.5–1.5%). After removing the essential oil, the peroxide values of the conventional method were very close to those of the NMR method. This indicated that the NMR method could be a more reliable method than the conventional method when the oil contains a significant amount of essential oil.

Table 3.2. shows the summary of different studies using the ^{1}H NMR spectroscopy for the oxidation of oil under storage conditions or accelerated storage conditions. Since the

Table 3.2 ¹H NMR spectroscopy used for oil oxidation under storage conditions

Oil	Temp. (°C)	Measurement	Conventional methods compared	Reference
Pollock liver oil, Sardine oil, methyl esters of pollock lover oil and sardine oil	40	Ratio of olefinic protons to aliphatic protons (0.5–3.0 ppm) (Ro)	Peroxide value (good correlation)	Saito (1987)
Sardine, saury and mackerel oils	40	Ratio of olefinic protons to aliphatic protons (0.5–3.0 ppm) (Ro)	Peroxide value (good correlation, NMR method can be used for a longer storage time)	Saito and Nakamura (1990)
Pollock oil, sardine oil	40	Ratios of olefinic protons (Ro) and diallylmethylene protons (Rm) to aliphatic protons (0.5–2.5 ppm)	Peroxide value (good correlation)	Saito and Nakamura (1989)
Sardine press cake	20	Ratios of olefinic protons (Ro) and diallylmethylene protons (Rm) relative to aliphatic protons (0.5–2.5 ppm)	Peroxide value, acid value, and carbonyl value (NMR method can be used for a longer storage time)	Saito and Udagawa (1992)
Canola oil, soybean oil	65	Ratios of aliphatic protons to olefinic protons (Rao) and to diallylmethylene protons (Rad)	TOTOX value (good correlation)	Wanasundara and Shahidi (1993)
Seal blubber oil, cod liver oil	65	Ratios of aliphatic protons to olefinic protons (Rao) and to diallylmethylene protons (Rad)	TOTOX value (good correlation), peroxide value, TBARS	Shahidi et al. (1994)
Canola oil, soybean oil	65	Ratios of aliphatic protons to olefinic protons (Rao) and to diallylmethylene protons (Rad)	TOTOX value (good correlation), peroxide value, TBARS (NMR method can be used for a longer storage time)	Wanasundara et al. 1995)
Sesame oil	65	Ratios of aliphatic protons to olefinic protons (Rao) and to diallylmethylene protons (Rad)	NMR method was used along with other conventional methods	Abou-Gharbia et al. (1996); Abou-Gharbia et al. (1997)

(continued)

Table 3.2 (continued)

Oil	Temp. (°C)	Measurement	Conventional methods compared	Reference
Borage oil, evening primrose oil	60	Ratios of aliphatic protons to olefinic protons (Rao) and to diallylmethylene protons (Rad)	Conjugated diene value, TBARS (good correlations)	Senanayake and Shahidi (1999)
Ethyl ester of docosahexaenoic acid	25	Multiple peaks at 8–10.5 ppm	Peroxide value, conjugated diene value, TBARS (good correlations)	Falch et al. (2004)
Capelin, cod liver, and tuna oils, blend of fish oils.	Room temp and 40	Fatty acid composition, ratios of aliphatic protons to olefinic protons (Rao) and to diallylmethylene protons (Rad)	Gas chromatography (good agreement), Peroxide value (good correlation)	Tyl et al. (2008)
Soybean oil	60	Fatty acid composition	Peroxide value, TBARS (good agreement)	Wang et al. (2014)
A variety of edible oils	40	Signal of hydroperoxide (OOH): used $CDCl_3$ and DMSO-d_6 (5:1, v:v) as solvent	Peroxide value (good agreement in general)	Skiera et al. (2012a)

first report by Saito in 1987, all the studies reported that the ^{1}H NMR method monitoring changes of a few specific signals were very reliable for the assessment of lipid oxidation during storage and had an advantage over conventional methods that it can be used for a much longer storage time. Since the NMR method simultaneously measures both the primary and the secondary oxidative changes in oils, it is more reliable than other methods measuring one or a few kinds of oxidation products. Therefore, NMR spectroscopy is considered a more suitable means for estimating lipid oxidation than chemical determinations (Shahidi and Zhong 2005). In addition, compared to the GC method, which is the most widely used method for analysis of fatty acid composition of oil, the ^{1}H NMR spectroscopy is a faster and nondestructive method for detection of the changes of the fatty acid profile (Pignitter and Somoza 2012).

3.2 Assessment of Oxidation During Frying

Deep fat frying is a cooking method, in which foods are submerged in extremely hot oil (typically at 160–190 °C). Deep fat frying is widely used due to the good taste and odor of resulting fried foods, the crispness and brown color on the outside of food, and the shorter cooking time. However, the frying process at such a high temperature can deteriorate the nutritional value of food and oil much faster than other cooking methods. Additionally, it can build up toxic compounds in oil, especially when the oil is repeatedly used. Therefore, it is rather important to ensure the oil quality during frying. Most countries established a regulatory limit of 25% total polar compounds in oil for human health (Dobarganes 2009; Rossell 2001).

Some above-mentioned analytical methods are not suitable to determine the oxidation level of frying oil. For example, hydroperoxides, the primary oxidation products, tend to decompose at the frying temperature and therefore, using the peroxide value for oxidative deterioration of frying oil is problematic (Augustin and Berry 1983; Farhoosh and Moosavi 2009). Conjugated diene value, total polar compounds, polymerized triacylglycerols, and carbonyl value were found to be relatively reliable indicators for oil oxidation during frying (Farhoosh and Tavassoli-Kafrani 2011). The gas chromatography (GC) determining the amounts of fatty acids in oil was also known to be a very reliable method showing strong correlations with organoleptic scores (Blumenthal et al. 1976). However, these methods also have certain problems as mentioned in Sects. 1.2 and 1.3.

^{1}H NMR methods developed for analysis of lipid oxidation can be divided into four different categories: (1) methods monitoring the changes in intensity of major NMR signals such as olefinic, allylic and bisallylic (= diallylmethylene) proton signals, (2) methods using the NMR proton relaxation time, (3) methods determining molar percentages of specific acyl groups and iodine value, and (4) methods measuring the amounts of oxidation products such as aldehydes, diglycerides, and epoxides. These four different types of NMR methods were studied for frying oils or oils oxidized at a frying temperature.

3.2.1 Methods Monitoring the Changes of Major NMR Signals

As described in Sect. 3.1, this method was developed to monitor oil oxidation under storage conditions and showed great potential as a new analytical method (Saito 1987). However, it was not utilized for frying oils for a while. Claxson et al. (1994) used ^{1}H NMR spectroscopy to find oxidation products produced during the oxidation of culinary oils and fats at 180 °C. Although the report focused on the analysis and identification of oxidation products generated, the authors also briefly discussed that the NMR analysis could follow the thermally induced consumption of polyunsaturated fatty acids (PUFAs) by quantifying reductions in the intensities of allylic (2.06 ppm), bisallylic (2.76 ppm), and olefinic protons (5.38 ppm). Khatoon and Krishna (1998) analyzed the ^{1}H NMR spectra of safflower oil samples heated at 180 °C for 10, 15, 90, 180, 360, and 480 min and observed that the olefinic proton signal at 5.4 ppm and the allylic proton signal at 2.85 ppm were already completely disappeared in 10 min and therefore, the gradual signal decrease could not be followed in this study. This fast disappearance of ^{1}H NMR signals was caused by the accelerated heating conditions (500 mL oil in a 26.5 cm diameter oven: surface area of 551.8 cm^2). This study reported the appearance of O–CH$_2$ or O–CH signal around 4 ppm in the ^{1}H NMR spectrum. At 90-min heating, a new peak at 6.5 ppm corresponding to the OH group appeared, which disappeared upon further heating. The formation of the OH group was also confirmed by the Fourier transformed infrared (FT-IR) spectrum. The CH$_2$ signal at 1–2 ppm and the CH$_3$ signal at 0.9 ppm became broader as the oil was heated, indicating that the original oil molecules were transformed to a variety of different oxidation products.

In 2006, nineteen years after the Saito's NMR method (Saito 1987) was reported, this method was first utilized to determine oil oxidation occurred at a frying temperature (150 °C) (Valdés and Garcia 2006). This research group analyzed ^{1}H NMR spectra of olive oil, sunflower oil, and the 1:1 mixture of olive oil and sunflower oil after oxidizing them at 150 °C. It was found that the relative intensities of olefinic, bisallylic, and allylic protons decreased, and those of α-CH$_2$ and β-CH$_2$ of the carboxyl group increased as the oxidation proceeded. In 2012, Hwang et al. (2012) practically used ^{1}H NMR spectroscopy to determine oil oxidation at a frying temperature (180 °C). The ^{1}H NMR method was slightly modified from the previous methods and used in a study on antioxidant activities of lignans and sesamol added in soybean oil. Figure 3.1 shows the ^{1}H NMR spectrum of soybean oil. In this study, individual peaks were integrated using the glycerol backbone CH$_2$ protons

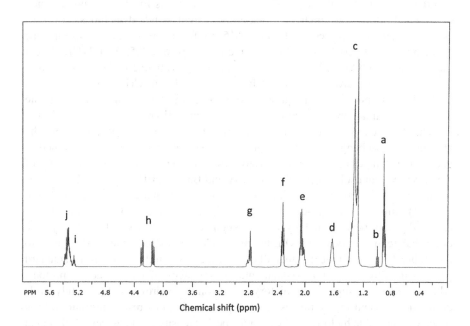

Fig. 3.1 ^{1}H NMR spectrum of soybean oil and peak assignments. (*a*) terminal CH$_3$, (*b*) terminal CH$_3$ of linonenic, (*c*) CH$_2$ other than CH$_2$ noted elsewhere, (*d*) (C=O)CH$_2$**CH$_2$**, (*e*) allylic protons, –**CH$_2$**CH=CH–, f: (C=O)**CH$_2$**–, g: bisallylic protons, –CH=CHCH$_2$CH=CH–, h: glycerol backbone CH$_2$, (*i*) glycerol backbone CH, (*j*) olefinic protons, –**CH=CH**–. From Hwang et al. (2012), used with permission

(peak h, the two double doublets, four protons, at 4.06–4.36 ppm) as the standard peak, instead of the aliphatic proton signal at 0.5–3.0 ppm (Saito 1987) and 0.6–2.5 ppm (Saito and Nakamura 1989) in the previously reported methods. The changes of the olefinic proton peak (peak j) intensity and the bisallylic peak (peak g) intensity were very similar to the trend of polymerized triacylglycerols (PTAG). The ¹H NMR and ¹³C NMR spectra revealed that there were no noticeable hydrolysis products (free acid: 10–12 ppm in ¹H NMR, 180 ppm in ¹³C NMR) or peroxides (8.2–8.9 ppm in ¹H NMR). Meanwhile, peaks of conjugated dienes and aldehydes were observed at 5.47–6.58 ppm and 9.4–9.8 ppm, respectively, in the ¹H NMR, which were previously reported by other research groups (Guillén and Ruiz 2004; Wanasundara et al. 1995).

One major convenience of the NMR method is that there is no need to use accurate amounts of the sample and the solvent. In the typical procedure of the NMR method, approximate amounts of the sample and the solvent are added in the NMR tube. This is possible because the intensity of a certain signal relative to a standard peak, such as the aliphatic proton signal or the glycerol backbone CH_2 signal, is measured from the NMR spectrum. Therefore, it eliminates the problems of other analytical methods such as sample-to-sample variations in concentration from inevitable human errors in weighing and pipetting resulting in very high repeatability and reproducibility. The procedure requires about 1 min for the sample preparation, less than 10 min for data acquisition, and a few minutes for data analysis, which offers much more convenience compared to current standard methods. The suggested experimental procedure (Hwang 2015) for frying oil analysis is as simple as: (1) Approximately 20–80 mg of oil sample is dissolved in 0.5–0.6 mL $CDCl_3$ and transferred in an NMR tube; (2) the NMR spectrum is obtained; (3) peaks are integrated using a standard signal such as the glycerol backbone CH_2 signal. Nowadays, an NMR instrument is often equipped with an auto-sampler, which can eliminate the need for human presence during many data acquisitions.

While the relative signal changes are used for the routine analysis of oil, the absolute signal changes in soybean oil during heat at 180 °C were determined to understand changes in non-reactive protons in soybean oil, using known amounts of the oil sample and solvent ($CDCl_3$) (Hwang and Bakota 2015). In this study, it was found that several proton peaks including peaks of terminal CH_3 groups (peak a in Fig. 3.1), aliphatic CH_2 groups (peak c), CH_2 groups at α- and β-positions to the carbonyl group (peak e), and glycerol backbone CH_2 groups (peak h) had very little or no change in their peak areas during the accelerated oxidation at 180 °C for up to 8 h (Table 3.3). This indicated that all of these signals could be used as the standard peak to monitor the relative signal intensity changes of other reactive protons. Although there are some contradictory reports on the hydrolysis of triacylglycerols during heating and frying processes of vegetable oils, this research group did not observe measurable hydrolysis affecting the peak intensity of the glycerol CH_2 signal (Hwang 2012, 2013b, 2017).

This ¹H NMR method that measures the changes of olefinic protons and bisallylic protons relative to the glycerol CH_2 signal to monitor oil oxidation was also utilized in an actual frying process, in which potato pieces were fried in soybean oil

Table 3.3 Percent area changes of NMR signals of soybean oil after 8-h heating at 180 °C in an accelerated oxidation test (Hwang and Bakota 2015), used with permission

NMR signals	Chemical shift, ppm	% Area change
Terminal CH_3	0.84–0.95	0.1
α-Linolenic, terminal CH_3	0.95–1.01	−9.6
$CH_3(CH_2)_n-$	1.10–1.50	0.3
$-COCH_2CH_2$	1.53–1.72	2.9
$-CH_2CH_2CH=CH$	1.72–1.93	−5.0
$-COCH_2$	2.25–2.39	0.2
$-CH=CHCH_2CH=CH-$	2.71–2.85	−11.7
Glycerol CH_2	4.06–4.36	0.1
Glycerol CH	5.23–5.30	0.1
$-CH=CH-$	5.30–5.45	−8.0

with small-scale frying equipment at 180 °C (Hwang et al. 2013b). As observed in the heating study, it was found that olefinic (5.29–5.45 ppm), bisallylic (2.17–2.85 ppm) and allylic (1.91–2.11 ppm) proton signals showed very similar trends to the polymerized triacylglycerols (PTAG) values determined by GPC analysis. The same ^1H NMR method was used to determine antioxidant activities of different steryl ferulates in a frying study, in which potato chips were fried in soybean oil at 180 °C (Winkler-Moser et al. 2015). Three other separate heating studies with soybean oil at 180 °C (Hwang and Winkler-Moser 2014; Hwang et al. 2014; Winkler-Moser et al. 2013) also showed that the ^1H NMR method was a very convenient analytical method to determine the oxidation level of soybean oil.

In a separate study, soybean oil and oils rich in oleic acid, mid-oleic sunflower oil (NuSun), and high oleic soybean oil (HOSBO) were heated at 180 °C or used for frying of tortilla chips at 180 °C (Hwang et al. 2017). In this study it was observed that, as shown in the earlier study (Hwang and Bakota 2015), signals of terminal CH_3 groups, aliphatic CH_2 groups, CH_2 groups at α- and β-positions to the carbonyl group, and glycerol backbone CH_2 groups did not show significant changes (Hwang et al. 2017). This result confirmed that any of these non-reactive proton signals could be used as a standard signal for monitoring the relative proton signal changes. Among them, the glyceride backbone CH_2 (4.04–4.39 ppm) could be the best standard peak, simply because the range of 4.04–4.39 ppm is generally less crowded than other regions and will have a lower chance of interference with signals of other components in oil such as antioxidants.

In this study, a systematic study on the reliability of the ^1H NMR method was conducted, in which olefinic, bisallylic, allylic, and linolenic CH_3 signals were used to monitor the oil oxidation. Correlations of the ^1H NMR method with conventional methods, including total polar compounds (TPC), polymerized triacylglycerols (PTAG) and the loss of polyunsaturated fatty acids (PUFAs), were examined. These conventional methods are believed to be the most reliable analytical methods for studying the degradation of oils under frying conditions. The study used regular soybean oil, mid-oleic sunflower oil (NuSun), and high oleic soybean oil (HOSBO)

Table 3.4 Fatty acid compositions and NMR signals of soybean oil (SBO), mid-oleic sunflower oil (NuSun) and high oleic soybean oil (HOSBO) (Hwang et al. 2017)

Fatty acid composition by gas chromatography (%)			
Fatty acid	*SBO*	*NuSun*	*HOSBO*
C14:0	0.07	0.05	0.04
C16:0	10.46	4.60	6.79
C18:0	5.05	3.60	3.64
C18:1	23.40	66.70	79.24
C18:2	53.49	23.49	7.54
C18:3	6.83	0.34	2.15
C20:0	0.39	0.33	0.36
C22:0	0.31	0.90	0.24
NMR signals and their proton numbers relatives to glyceride CH$_2$			
Signals	*Chemical shift (ppm)*	*SBO*	*NuSun*	*HOSBO*
Olefinic H	5.16–5.30	8.93	6.88	6.00
Glyceride CH	5.30–5.46	1.00	1.00	1.00
Glyceride CH$_2$	4.04–4.39	4.00	4.00	4.00
Bisallylic CH$_2$	2.70–2.88	3.95	1.36	0.60
(C=O)CH$_2$	2.22–2.41	6.15	6.16	6.16
Allylic CH$_2$	1.94–2.15	10.22	11.19	10.91
Aliphatic CH$_2$	1.05–1.71	59.91	67.21	69.37
Terminal CH$_3$, linolenic	0.95–1.01	0.67	0.09	0.23
Terminal CH$_3$, others	0.74–0.95	8.59	9.23	9.00

heated at 180 °C. Table 3.4 shows fatty acid compositions and NMR signals of fresh oils used in this study.

Oleic acid (C18:1) oxidizes much slower than linoleic acid (18:2) and linolenic acid (18:3). Therefore, high oleic oils are recognized as the most promising alternatives to hydrogenated oils containing *trans* fats and/or highly saturated fats and tropical oils rich in saturated fats. For soybean oil containing 23.4% oleic acid, 53.5% linoleic acid, and 6.83% linolenic acid, changes of all four proton signals (olefinic, bisallylic, allylic, and linolenic CH$_3$ signals) relative to the glyceride backbone CH$_2$ signal showed strong correlations with conventional methods (TPC, PTAG, and concentrations of linoleic acid and linolenic acid determined by GC, $R^2 = 0.9704$–0.9917). In the experiments with NuSun consisting of 66.7% oleic acid, 23.5% linoleic acid, and 0.34% linolenic acid, olefinic proton and allylic proton signals relative to the glyceride backbone CH$_2$ signal showed good correlations with the conventional methods ($R^2 = 0.8915$–0.9820). The linolenic terminal CH$_3$ signal could not be used as an indication of oxidation for NuSun because the signal (0.09 protons) was too small. The bisallylic proton signal was also relatively small (1.36 protons) and showed relatively weak correlations with the conventional methods. HOSBO, containing 79.2% oleic acid, 7.5% linoleic acid, and 2.2% linolenic acid, showed strong correlations between relative changes in olefinic proton, allylic proton signals, and the conventional methods ($R^2 = 0.9185$–0.9844). Similar to NuSun, the linolenic CH$_3$ signal and the bisallylic proton signal were not recommended as the indication of oxidation due to the relatively smaller peaks (0.23 linolenic CH$_3$ protons and 0.60 bisallylic protons). This study provides important information indicating that the signals to monitor the deterioration of oil should be carefully selected, depending on the oil used. Olefinic proton and allylic proton signals are considered to be the most reliable signals for vegetable oils since they usually appear as strong signals in ^{1}H NMR spectrum.

3.2.2   Methods Using the NMR Proton Relaxation Time

In 1996, Sun and Moreira (1996) investigated if there were good correlations between NMR proton relaxation times and the increases in free fatty acids and polar materials in a study with the soybean oil samples oxidized by heating at 190 °C and by frying corn tortilla dough at 190 °C. The results showed that the longitudinal relaxation time (T_1) and the transverse relaxation time (T_2) showed a linear relationship with free fatty acids, giving strong correlations (R^2 = 0.985, 0.950, respectively). T_1 and T_2 also showed strong correlations with total polar materials (R^2 = 0.99, 0.95, respectively). The theory behind this method is that there are molecular structure and composition changes in oil during the oil degradation process, affecting the chemical environment surrounding the protons. Thus, the proton mobility affecting the NMR proton relaxation times changes as oil degrades. Another group also examined the correlation between the relaxation time in NMR spectroscopy and the total polar compounds (TPC) (Hein et al. 1998) in an attempt to develop a new analytical method to replace the time and chemical consuming TPC method. This group found the correlation coefficient (r) to be 0.931–0.944 between the NMR method and TPC in frying experiments with peanuts, snack products, cuttlefish, and French fries fried in palm oil at 190 °C. Based on findings by Sun and Moreira (1996) and Hein et al. (1998), structural and compositional changes in oil during the oil degradation process affect the NMR relaxation time. This means that the solid fat content (SFC) value, which is also determined by the relaxation time, is expected to reflect the oil degradation. Bakota et al. (2012) found that the SFC values of several plant oils including soybean, grapeseed, walnut, sunflower, and canola oils correlate positively with total polar compounds (TPC) and inversely with triglyceride concentration. The SFC analysis using the low-resolution NMR is a very convenient, relatively inexpensive, and user-friendly analytical method, and can be an alternative to the labor and time consuming total polar compound analysis.

3.2.3   Methods Measuring Acyl Groups

In a study conducted by Guillén and Uriarte (2009), they calculated the iodine values and molar percentages of acyl groups such as linoleic, oleic, and saturated acyl groups for sunflower oil heated at 190 °C using the equations developed earlier

(Guillén and Ruiz 2003b). They found that these values showed a linear trend with the heating time. This method was then utilized to monitor oxidation of linseed oil (Guillén and Uriarte 2012a), sunflower oil (Guillén and Uriarte 2012b), and extra virgin olive oil (Guillén and Uriarte 2012c) heated in an industrial fryer at 190 °C. In separate study (Guillén and Uriarte 2013), this research group developed an equation that can predict the percentage of polar compounds produced during heating extra virgin olive, sunflower, and virgin linseed oils at 190 °C using the molar percentages of triunsaturated, diunsaturated and monounsaturated acyl groups determined by ¹H NMR signals. This study revealed the close relationships between the empirical results of polar compounds and the predicted polar compounds calculated from the equation. During the actual deep-frying process with Spanish doughnuts, pork adipose tissue, and farmed salmon fillets in extra virgin olive, soybean, and sunflower oils, it was confirmed that the iodine value and molar percentages of acyl groups calculated from the NMR spectrum could be a good indication of oil oxidation (Martínez-Yusta and Guillén 2014a, b). Another group used ¹H NMR data to determine the molar percentages of main acyl groups in the frying oils and fish lipids during shallow-frying of fillets of farmed gilthead sea bream (*Sparus aurata*) and European sea bass (*Dicentrarchus labrax*) in extra-virgin olive and sunflower oils (Nieva-Echevarría et al. 2016). The study confirmed that there was a migration of main and some minor compounds between oil and fish. The same technique was used to monitor the evolution of the compositional changes in sunflower oils throughout fourteen deep-frying experiments of three different foods (Martínez-Yusta and Guillén 2016). For a heating process, the changes in the molar percentages of acyl groups in sunflower oil are caused by thermo–oxidation. However, for the frying process involving foods, the migration of lipid from the food to the frying medium significantly affected the molar percentage of all kinds of acyl groups. The NMR technique could also detect the formation of the genotoxic and cytotoxic 4-hydroxy-(E)-2-alkenals and 4,5-epoxy-2-alkenals, along with other aldehydes. In addition, it was found that proton signals of 1,2-diglycerides and 1,3-diglycerides) increased their intensity with the processing time.

3.2.4 Methods Measuring Aldehydes and Other Oxidation Products

The ¹H NMR was also used to analyze aldehydes produced from palm, sunflower, and soybean oils heated or used for frying potatoes at 180 °C (Romano et al. 2009). From the investigation of the ¹H NMR spectra of oxidized oils, it was confirmed that n-alkanals, 4-hydroxy-*trans*-allenals, alka-2,4-dienals, and *trans*-2-alkenals were accumulated, and that 2-alkenals and alka-2,4-dienals are the major aldehydes formed during heating or frying. When Guillén and Uriarte (2009, 2012b) conducted studies on the NMR methods measuring acyl groups and iodine values, they also analyzed a variety of aldehydes from the NMR spectra of sunflower oil heated at 190 °C. They found that the intensity of the aldehydic protons signal (CHO) of aldehydes including n-alkanals (9.748 ppm, triplet), (E)-2-alkenals (9.493 ppm, doublet), (E,E)-2,4-alkadienals (9.520, doublet), (E,Z)-2,4-alkadienals (9.593, doublet), 4-hydroxy-(E)-2-alkenals (9.573, doublet), and 4-oxoalkanals (9.780, triplet) could be a good indication of oil degradation during heating at frying temperature. It was noted that concentrations of (E,E)-2,4-alkadienals, (E,Z)-2,4-alkadienals, and 4-hydroxy-(E)-2-alkenals reached a maximum value and then decreased during the heating process. In contrast, concentrations of (E)-2-alkenals, n-alkanals, and 4-oxoalkanals increased during the whole period of heating time (40 h). The NMR signals of hydroperoxide protons were not observed in this study and the peroxide value of the oil remained almost constant throughout the oil degradation process. Very similar trends in the formation of these aldehydes were also observed with linseed oil heated at 190 °C for 20 h (Guillén and Uriarte 2012a), extra virgin olive, sunflower, and virgin linseed oils (Guillén and Uriarte 2012c) during heating at a frying temperature (190 °C). In the study with extra virgin olive, sunflower, and virgin linseed oils (Guillén and Uriarte 2012c), the formation of (E) and (Z)-9,10-epoxystearyl groups was also observed in addition to aldehydes, and their concentrations showed linear relationships with the heating time. It should be noted that oils with the same amount of polar compounds had different safety levels with different compositions of oxidation products including genotoxic and cytotoxic 4-hydroxy-alkenals, although the safety level of the oxidized oil is currently judged by the amount of polar compounds in many countries.

Nieva-Echevarría et al. (2016) measured the amounts of 1,2-diglycerides and aldehydes in extra-virgin olive oil and sunflower oil in which fish was fried. It seemed that the amount of 1,2-diglyceride would not be a good indication of oil oxidation, as its change was small and not consistent in this study. However, in another study, intensities of NMR signals of primary alcohols and 1,2-diglycerides constantly increased with the processing time during the frying of Spanish doughnuts, pork adipose tissue, and salmon fillets in sunflower oil (Martínez-Yusta and Guillén 2016). It has been controversial whether the hydrolysis of triacylglycerols or the amount of free fatty acids is a good indication of lipid oxidation.

Table 3.5 shows the summary on the NMR methods reported by many research groups. Since the research groups that used the methods measuring acyl groups also studied the occurrence of oxidation products such as aldehydes, diglycerides, and epoxides, these two methods are combined in the table.

Table 3.5 ¹H NMR spectroscopy for oil oxidation at frying temperatures or under frying conditions

Oil	Conditions	Measurement	Comparison with conventional methods	Reprehensive references
Methods monitoring the intensity of major peaks				
Vegetable oils, lard, ghee	Heating in glass vessels at 180 °C	Intensity of allylic, bisallylic, and olefinic protons	None	Claxson et al. (1994)
Safflower oil	Heated in an open pan at 180 °C	Observed changes in NMR signals including olefinic and allylic proton signals	Free fatty acid, peroxide value, iodine value, fatty acid composition, conjugated diene and triene, color, viscosity, and IR were monitored along with the NMR signals	Khatoon and Krishna (1998)
Olive and sunflower oils	Heated at 150 and 225 °C	Intensities of olefinic, bisallylic, and allylic protons	Density, interfacial tension, and degree of unsaturation were monitored along with the NMR signals	Valdés and Garcia (2006)
Soybean oil	Heating at 180 °C	Olefinic, allylic , and bisallylic proton signals	Observed similar trends with the amounts of polymerized triacylglycerols	Hwang et al. (2012)
Soybean oil	Frying at 180 °C	Olefinic, allylic , and bisallylic proton signals	Observed similar trends with the amounts of polymerized triacylglycerols	Hwang et al. (2013b), Winkler-Moser et al. (2015)
Methods using the NMR proton relaxation time				
Soybean oil	Heating and frying at 190 °C	NMR proton relaxation times (T_1 and T_2)	Linear relationship with the increase in free fatty acids and polar materials	Sun and Moreira (1996)
Palm oil	Frying at 190 °C	NMR proton relaxation times	Good correlation with total polar compounds	Hein et al. (1998)
Soybean, grapeseed, walnut, sunflower, canola oils	Heated in glass NMR tubes at 180 °C	The low-resolution NMR used for solid fat content (SFC)	Good correlation with total polar compounds	Bakota et al. (2012)

(continued)

Table 3.5 (continued)

Oil	Conditions	Measurement	Comparison with conventional methods	Reprehensive references
Methods measuring acyl groups, iodine value, aldehydes, and other oxidation products				
Palm, sunflower, and soybean oils	Frying at 180 °C	Aldehydes	None	Romano et al. (2009)
Sunflower oil	Heating at 190 °C	Acyl groups iodine values, and aldehydes	None	Guillén and Uriarte (2009), Guillén and Uriarte (2012b)
Sunflower oil	Frying at 190 °C	Acyl groups, iodine values, diglycerides, aldehydes, and alcohol	Used along with polar compounds measured by Testo 265 instrument	Martínez-Yusta and Guillén (2016)
Linseed oils	Heating at 190 °C	Acyl groups iodine values, and aldehydes	Observed similar trends with iodine value and polar compounds measured by Testo 265 instrument.	Guillén and Uriarte (2012a),
Olive oil	Heating at 190 °C	Acyl groups iodine values, and aldehydes	Good correlation with polar compounds measured by Testo 265 instrument	Guillén and Uriarte (2012c)
Olive, sunflower and linseed oils	Heating at 190 °C	Acyl groups iodine values, and aldehydes	Good correlation between predicted polar compounds and experimental polar compounds measured by Testo 265 instrument	Guillén and Uriarte (2013)
Olive and sunflower oils	Shallow-frying of fish in a pan at 170 °C and microwave-frying at 900 W	Acyl groups, aldehydes, and diglycerides	None	Nieva-Echevarría et al. (2016)

Chapter 4
^{1}H NMR Spectroscopy for Identification of Oxidation Products and for Elucidation of Reaction Mechanisms

In addition to the assessment of lipid oxidation, one of the most significant contributions of the ^{1}H NMR technique is that it made major advances in elucidation of molecular structures of oxidation products. This led to a better understanding of lipid oxidation mechanisms. It is almost impossible to fully understand the complicated oxidation mechanisms and indentify all of the oxidation products. However, the ^{1}H NMR technique has been the most powerful tool for this purpose, along with some other NMR techniques such as ^{13}C, and ^{31}P NMR and their 2D NMR techniques. The method to identify oxidation products also can be used to assess the quality of edible oils by checking the appearance of specific oxidation products, such as aldehydes and peroxides, in addition to the aforementioned methods to determine lipid oxidation (Chakraborty and Joseph 2016).

4.1 ^{1}H NMR to Indentify Oxidation Products During Storage of Oil

Some oxidation products are easily decomposed and not found in oils oxidized at high temperatures such as frying temperatures. With a greater variety of oxidation products identified, better understanding of oxidation mechanisms could be achieved. For this reason, early studies were conducted at relatively lower temperatures such as storage temperatures or slightly higher temperatures for the acceleration of oxidation (25–70 °C). And of course, later, it was found that there were significant differences between oxidation products and oxidation mechanisms with different oxidation temperatures, which will be discussed in the next section.

Nowadays, high resolution NMR instruments (typically 300–600 MHz) are popularly used. However, early low resolution NMR devices (60–100 MHz) also made significant contributions in the identification of oxidation products. In 1966, Zimmerman

© The Author(s) 2017
H.-S. Hwang, *Advances in NMR Spectroscopy for Lipid Oxidation Assessment*,
SpringerBriefs in Food, Health, and Nutrition, DOI 10.1007/978-3-319-54196-9_4

(1966) utilized a very early model of the NMR instrument (Varian A-60 A, 60 MHz) to verify the isomerization of unsaturated hydroperoxide to a ketohydroxy compound. In 1973, Pokorný et al. (1976) used a Varian LX-100 apparatus (100 MHz) and reported that the methine proton in >C<u>H</u>COOH group at 5.36 ppm evidenced the formation of peroxides during oxidation of methyl esters of olive oil at 60 °C, In 1977, in an attempt to indentify oxidation products from methyl linoleate, Chan and Levett (1977) successfully elucidated the molecular structures of four major oxidation products using Perkin Elmer R32 (90 MHz) spectrometer. This research group separated the four major oxidation products of methyl linoleate oxidized at 30 °C by HPLC, and then compared their NMR spectra with known hydroperoxides. The four compounds were identified as 13-*L*-hydroperoxy-*cis*-9-*trans*-11-octodecadienoate, 9-*D*-hydroperoxy-*trans*-10-*cis*-12-octadecadienboate, 13-hydroperoxy-*trans,trans*-9,11-octadecadienoate, and 9-hydroperoxy-*trans,trans*-10,12-octadecadienoate (Fig. 4.1).

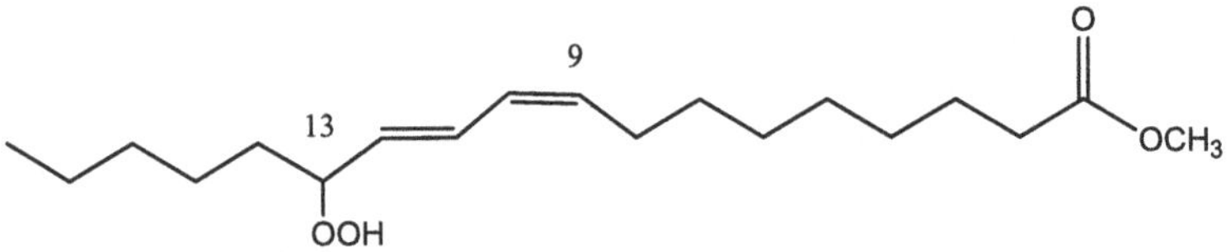

13-*L*-hydroperoxy-*cis*-9-*trans*-11-octodecadienoate

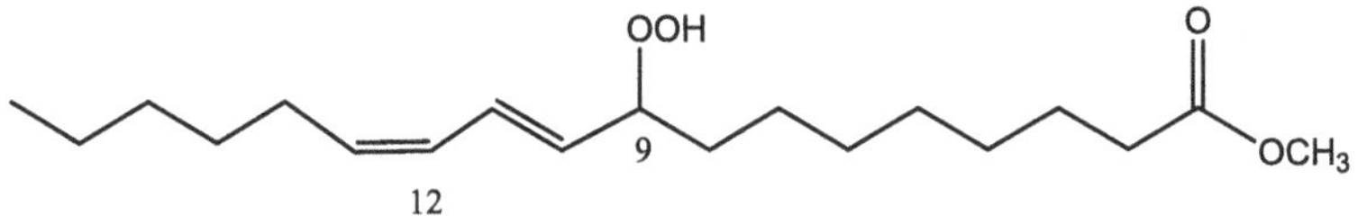

9-*D*-hydroperoxy-*trans*-10-*cis*-12-octadecadienboate

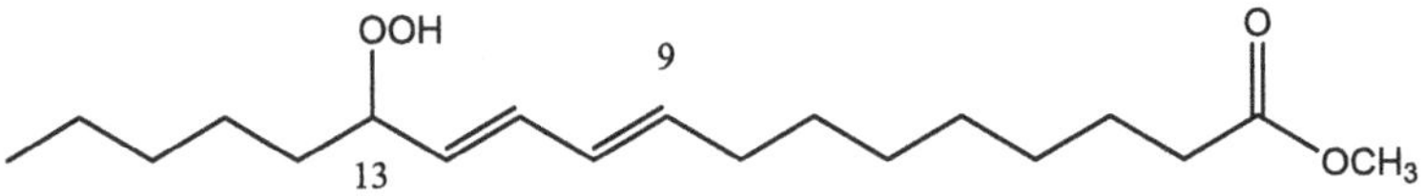

13-hydroperoxy-*trans,trans*-9,11-octadecadienoate

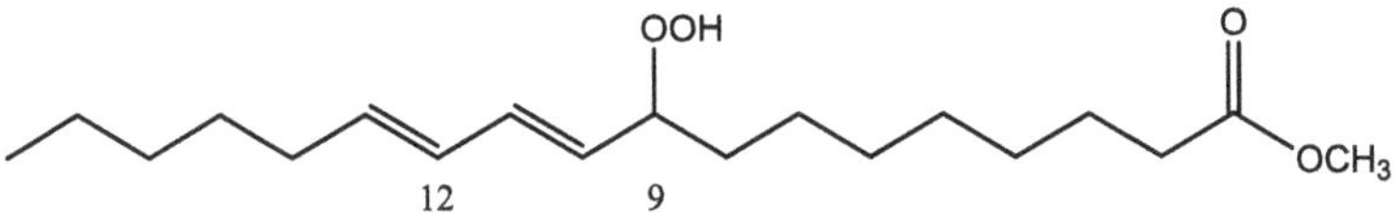

9-hydroperoxy-*trans,trans*-10,12-octadecadienoate

Fig. 4.1 Four hydroperoxides identified by Chan and Levett (1977)

Frankel et al. (1982) used an NMR instrument (100 MHz) to characterize oxidation products from photosensitized oxidation of methyl linoleate. Photosensitized oxidation was conducted by dissolving methyl linoleate in CH_2Cl_2 containing 10 mg methylene blue/g of methyl linoleate, and exposing to a 1000 W light at 0 °C. They found greater concentrations of 9- and 13-hydroperoxides than 10- and 12-hydroperoxides under these conditions. They also found four hydroperoxy cyclic peroxides (Fig. 4.2), which subsequently produce the major volatiles, hexanal and methyl 10-oxo-8-decenoate, by thermal decomposition. This research group proposed the mechanism for cyclization of internal 10-hydroperoxide (Fig. 4.3).

Further studies on photosensitized oxidation of methyl linoleate by Neff et al. (1983) found 6-membered hydroperoxy cyclic peroxides using a 300 MHz NMR instrument, along with ultraviolet, infrared, and mass spectrometry. The 6-membered hydroperoxy cyclic peroxides were identified as 13-hydroperoxy-9,12-epidioxy-10- and 9-hydroperoxy-10,13-epidioxy-11-octadecenoates (Fig. 4.4). They also found two other keto diene products, 9-oxo-*trans,trans*-10-12- and 13-oxo-*trans,trans,*-9,11-octadecadienoates.

Guillen and Ruiz (2004) used ¹H NMR spectroscopy to identify oxidation products from sesame oil heated at 70 °C. They found a variety of aldehydes including 4-hydroxy-*trans*-2-alkenal, 4-hydroperoxy-*trans*-2-alkenal, and 4,5-epoxy-*trans*-2-alkenal. In turn, they also found more aldehydes including *trans-*

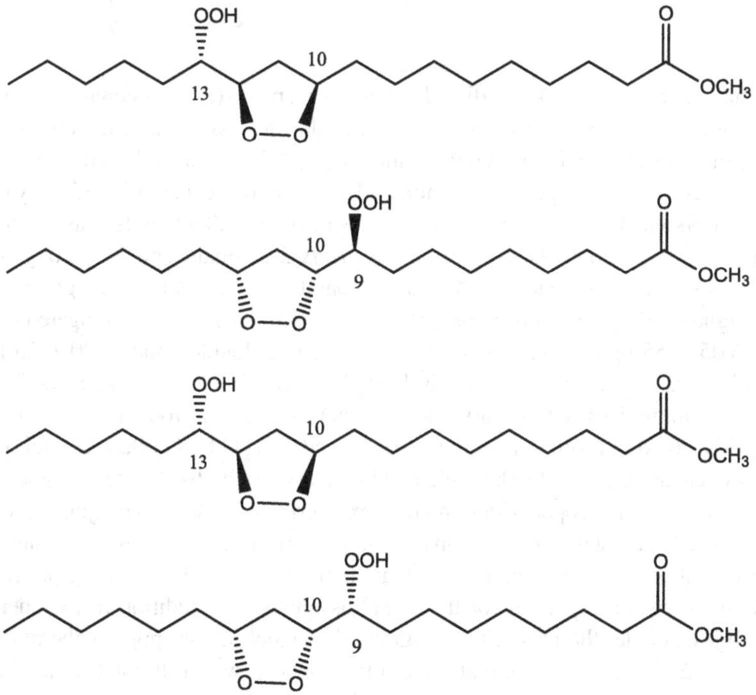

Fig. 4.2 Hydroperoxy cyclic peroxides from photosensitized oxidation of methyl linoleate (Frankel et al. 1982)

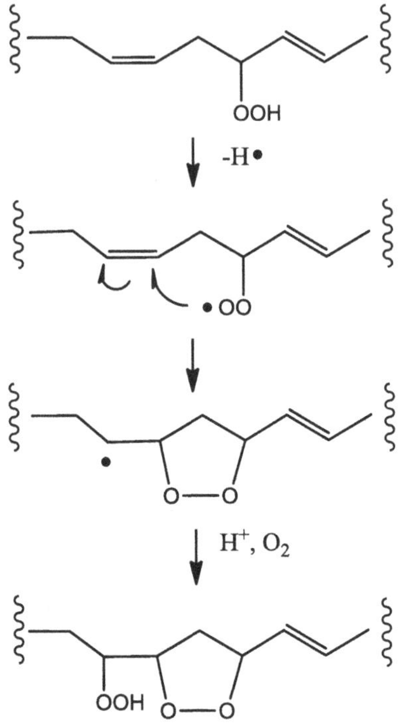

Fig. 4.3 Mechanism for
cyclization of internal
10-hydroperoxided
(Frankel et al. 1982)

2-alkenals, *trans,trans*-2,4-alkadienals, *trans*-4,5-epoxy-(*E*)-2-decenal, 4-hydroxy-*trans*-2-alkenals, 4-hydroperoxy-*trans*-2-alkenals, and *n*-alkanals from corn oil and sunflower oil heated at 70 °C (Guillén and Ruiz 2005b). In a study with oils rich in linolenic acid, such as rapeseed, walnut, and linseed oils heated at 70 °C, they could detect n-alkanals, *trans*-2-alkenals, *trans,trans*-2,4-alkadienals, as well as 4,5-epoxy-*trans*-2-alkenals, 4-hydroperoxy-*trans*-2-alkenals, and 4-hydroxy-*trans*-2-alkenals (Guillén and Ruiz 2005a). In a separate study, this research group could detect signals indicating hydroperoxides (8.3–8.6 ppm), a *cis, trans*-conjugated double bond (6.05–6.55 ppm), and a *trans, trans*-conjugated double bond (5.70–6.25 ppm) from olive, hazelnut, and peanut oils heated at 70 °C (Guillén and Ruiz 2005c). Table 4.1 summarizes the primary and secondary oxidation products observed from edible oils stored at room temperature, 70 °C, and 100 °C by Guillen research group (Goicoechea and Guillen 2010; Guillen and Goicoechea 2009). As seen in other studies, they observed hydroperoxides and hydroxy derivatives of (*Z,E*) conjugated-dienic systems at the early stage of oxidation and then, a variety of aldehydes at the later stage of the oxidation. As shown in Table 4.1, it should be noted that splitting patterns of NMR signals were important for the signal assignment in addition to the chemical shifts. For example, the reason for assigning the signal at 9.49 ppm as the aldehyde proton of (*E*)-2-alkenals was that its splitting pattern was a doublet indicating an adjacent CH group. On the other hand, the triplet signal at 9.75 indicates the adjacent CH₂ group and was identified as the aldehyde peak of alkanals.

Since ¹H NMR signals of oxidation products were well assigned, these signals can often be used in the determination of the quality and the level of oxidation of

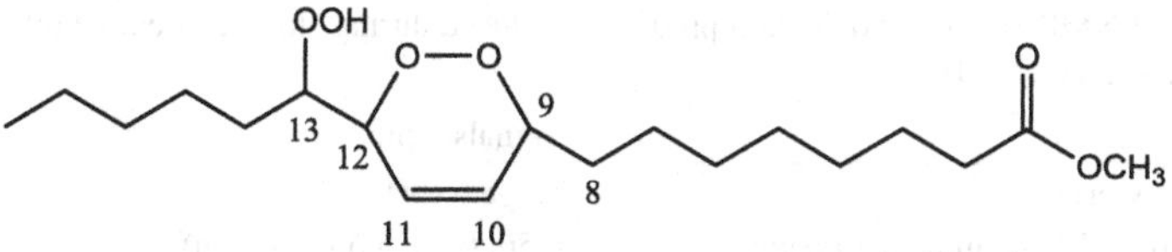

Methyl 13-hydroperoxy-9,12-epidioxy-10-octadecenoate

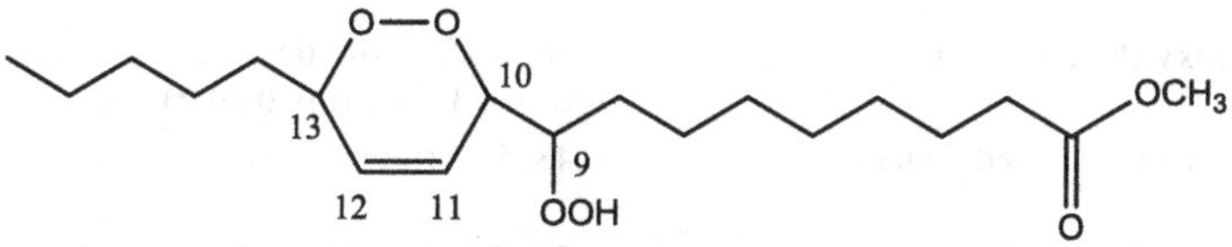

Methyl 9-hydroperoxy-10,13-epidioxy-11-octadecenoate

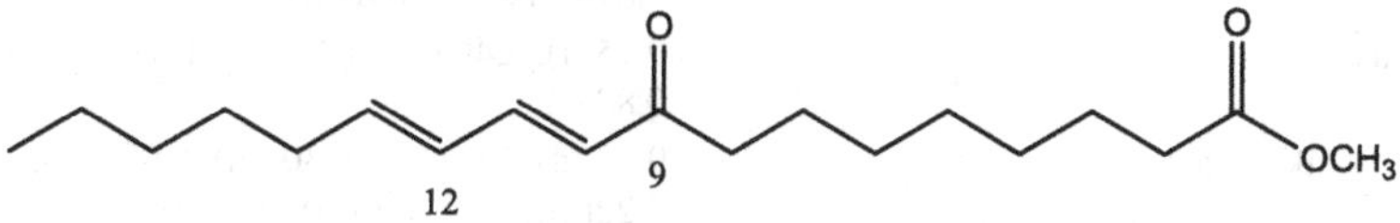

Methyl 9-oxo-*trans,trans*-10-12-octadecadienoate

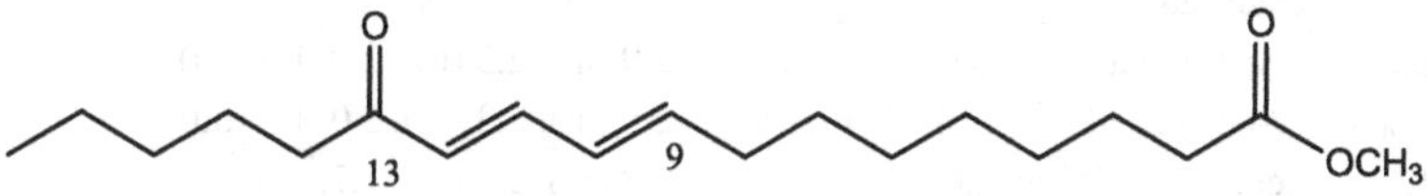

Methyl 13-oxo-*trans,trans*-9,11-octadecadienoates

Fig. 4.4 Structures of 6-memebred hydroperoxy cyclic peroxides and keto dienes produced by photosensitized oxidation of methyl linoleate (Neff et al. (1983)

edible oils. For example, the proton peaks of monohydroperoxide (broad singlet at 8.0–9.0 ppm) and conjugated dienes (multiplets at 5.5–6.6 ppm) were used to evaluate corn-zein edible coating, which was used to protect macadamia nuts from oxidation (Colzato et al. 2011). Signals of *Z,E-* and *E,E*-conjugated peroxides (5.4–6.7 ppm) and four different aldehyde products (9.4–9.8 ppm) were used to evaluate antioxidant activities of several compounds, including 4-vinylsyringol in soybean oil (Wang et al. 2014). Hydroperoxides, conjugated dienes, and aldehydes were monitored to study the antioxidant activity of caffeic acid phenethyl ester (Chakraborty and Joseph 2016).

4.2   ¹H NMR to Indentify Oxidation Products at Frying Temperatures

As mentioned in the previous sections, one major difference between the oxidation products formed at frying temperatures and relatively lower temperatures is that, in general, intermediate oxidation products such as hydroperoxides are not observed at

Table 4.1 ¹H NMR signals of oxidation products produced during storage of edible oils (Guillen and Goicoechea 2009, 2010)

Oxidation products	Signals, ppm
Hydroperoxy groups	8.3–8.9 (br)[a]
Hydroperoxy-(Z,E)-conjugated-dienic systems	6.56, 6.00, 5.58, 5.45 (m)
Hydroperoxy-(E,E)-conjugated-dienic systems	6.25, 5.75 (m)
4-Hydroperoxy-(E)-2-alkenals	9.58 (d), 8.20 (d), 6.80 (dd), 6.33 (m), 4.66 (dd), 1.73–1.21 (m), 0.89 (t)
Hydroxy-(Z,E)-conjugated-dienic systems	6.48, 5.98 (m)
4-Hydroxy-(E)-2-alkenals	9.57 (d), 6.84 (dd), 6.33(ddd), 4.43 (m), 1.30–1.70 (m), 0.90 (t)
(E)-2-alkenals	9.49 (d), 6.85 (tt), 6.11 (dd), 2.32 (q) 1.69–1.19 (bs), 0.89 (t)
Alkanals	9.75 (t), 2.40 (dt), 1.61 (m), 1.32–1.27 (bs), 0.88 (t)
(E,E)-2,4-alkadienals	9.53 (d), 7.09 (m), 6.30 (m), 6.08 (dd), 2.22(m), 1.47–1.3 (m), 0.90 (t)
4,5-Epoxy-(E)-2-alkenals	9.54 (d), 6.56 (dd), 6.40 (dd), 3.33 (dd), 2.96 (td), 1.65-1.33 (m), 0.91 (t)
9,10-Epoxyoctadecanoate	2.9 (m), 1.5 (m)
9,10-Epoxyoctadecenoate (leukotoxin)	2.9 (m), 2.3 (m), 2.0, 1.5 (dd)
12,13-Epoxy-9-octadecenoate (isoleukotoxin)	2.9 (m), 2.3 (m), 2.0, 1.5 (dd)
9,10-12,13-Diepoxyoctadecanoate	3.1 (m), 2.9 (m), 1.7, 1.5
9,10-Dihydroxy-12-octadecenoate (leukotoxindiol)	3.4 (m)

[a]Splitting patterns are shown in a parenthesis: br (broad singlet), d (doublet), dd (double doublet), ddd (double of double doublet), t (triplet), td (triple doublet), m (multiplet)

frying temperatures because these intermediate oxidation products easily react with other compounds or decompose to produce secondary oxidation products. For example, no hydroperoxides were detected in the ¹H NMR spectrum of sunflower oil during heating at 190 °C (Guillén and Uriarte 2009), while they were observed when oils were heated at relatively lower temperatures (such as at 70 °C) (Guillén and Ruiz 2004, 2005a, b, c). The only exception occurred when the oil was heated for a relatively short time. In a heating study in which several oils, including corn, sunflower, soybean, rapeseed, peanut, grapeseed, and olive oils, were heated at 180 °C for a relatively short period of time (30–90 min), hydroperoxides of conjugated diene systems were observed in the ¹H NMR spectra (Claxson et al. 1994). A hydroperoxide proton singlet peak (–OO<u>H</u>) and a multiplet for –C<u>H</u>(OOH)– were also detected at 8.5–8.9 ppm and 4.35 ppm, respectively, indicating the presence of hydroperoxides. In addition to these signals, this research group observed conjugated diene olefinic proton multiplets appeared in the 5.4–6.7 ppm range, α,β-unsaturated aldehydes at 9.48, 9.52, and 9.63 ppm, and saturated aldehydes at 9.74 ppm. The total concentrations of saturated and α,β-unsaturated aldehydes were determined to be 1–20 and $2–30 \times 10^{-3}$ mol/kg, respectively. Based on these results, it was concluded that primary oxidation products were formed at the early stage of the oxidation process during the frying process, and then these intermediate products quickly reacted or decomposed to other secondary oxidation products. Therefore, regardless of oxidation temperatures, it is generally accepted that the lipid oxidation begins with the reaction between lipid free radicals (L˙) and oxygen to form peroxyl radicals (LOO˙), followed by reactions of peroxides to produce numerous secondary oxidation products. Final oxidation products at different temperatures are different because the primary oxidation products may take different reaction routes at different temperatures. Again, lipid oxidation is quite complicated and is not fully understood yet.

Khatoon and Krishna (1998) studied the oxidation of safflower oil under different oxidation conditions. From the oil subjected to static heating at 180 °C for 8 h in an open pan, it was obvious that alcohols were produced from the hydroperoxide breakdown, according to the NMR spectra as well as the IR spectra. Signals indicating O–CH or O–CH$_2$ protons at around 4 ppm in the ¹H NMR spectrum and the peaks at 3368 cm^{-1} the IR spectrum indicating the OH group, increased with time. The IR spectrum indicated that there were carbonyl compounds formed during the oxidation process, as the peak at 1708 cm^{-1} (carbonyl group) increased as the oxidation proceeded. The ¹H NMR signal of the terminal CH$_3$ of fatty acids (0.9 ppm) broadened with time, indicating the formation of a various oxidation products.

Romano et al. (2009) observed n-alkanals, *trans*-2-alkenals, 4-hydroxy-*trans*-alkenals, 4-*oxo*-alkenals, and alka-2,4-dienals in the region of 9.3–9.7 ppm of the ¹H NMR spectra in a blend of fractionated and deodorized palm, sunflower, and soybean oils, during the thermo-oxidization and frying at 180 °C for 4 and 40 h. Guillén and Uriarte (2009) identified several aldehydes produced during heating of sunflower oil at 190 °C for periods of 8 h per day over 4 days from ¹H NMR spectra. Aldehydes including n-alkanals, (*E*)-2-alkenals, (*E,E*)-2,4-alkadienals, (*E,Z*)-2,4-alkadienals, 4-hydroxy-(*E*)-2-alkenals, and 4-oxoalkanals were identified based on the chemical shifts and splitting patterns of the aldehydic proton signals (CHO).

Guillén and Uriarte (2012b) reported that hydrolysis producing free fatty acids and mono- and di-glycerides is generally negligible if there is no water added during heating at frying temperatures. A study with virgin linseed oil also showed that the CH$_2$–OH signal (3.72 ppm, doublet), belonging to 1,2 di-glycerides and monoglycerides (3.6 ppm), remained almost unchanged. No other new signals attributed to mono- or di-glycerides appeared during heating at 190 °C, indicating negligible hydrolysis (Guillén and Uriarte 2012a). However, a more thorough investigation showed that signals attributable to the protons of primary alcohols (triplet at 3.62 ppm, –CH$_2$OH–) and secondary alcohols (multiplets at 3.54–3.59 ppm, >CHOH) newly appeared in the ¹H NMR spectra of extra virgin oil oxidized by heating at 190 °C and by deep-frying of Spanish doughnuts, pork adipose tissue, and salmon fillets for 7.5 h at 190 °C (Martínez-Yusta and Guillén 2014a). The concentration of secondary alcohols was much higher than that of primary alcohols. In an earlier study with safflower oil subjected to 180 ± 2 °C, monohydroxy-fatty acids and polyhydroxy-fatty acids were detected by chromatographic methods, which were, in turn, confirmed by ¹H NMR spectroscopy (Shantha and Decker 1994).

Signals of epoxides at 2.6 and 2.9 ppm in the NMR spectrum were not observed in sunflower oil and virgin linseed oil heated at 190 °C up to 32 h (Guillén and Uriarte 2012b). However, extra virgin olive oil submitted to the same heating conditions showed the signals of (*Z*)-9,10-epoxystearyl and (*E*)-9,10-epoxystearyl groups (Guillén and Uriarte 2012a). This result indicated that the formation of epoxides under frying conditions depends on the kind of oil used. Guillén and Uriarte (2012c) further investigated epoxy compounds produced in extra virgin olive oil heated at 190 °C and identified five epoxides: (*Z*)-9,10-epoxystearate, (*E*)-9,10-epoxystearate, 9,10-epoxy-12-octadecenoate (leukotoxin), 12,13-epoxy-9-octadecenoate (isoleukotoxin), and 9,10-12,13-diepoxy-octadecanoate (Fig. 4.5)

Table 4.2 shows the summary of signals appeared in the ¹H NMR spectrum of oil oxidized at frying temperatures or during frying processes.

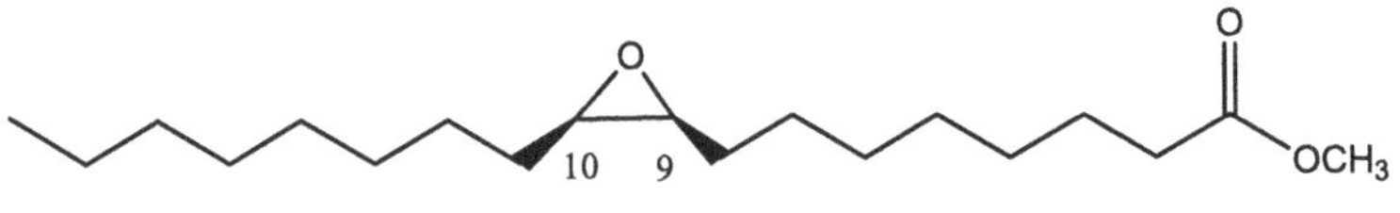

(*Z*)-9,10-Epoxystearate

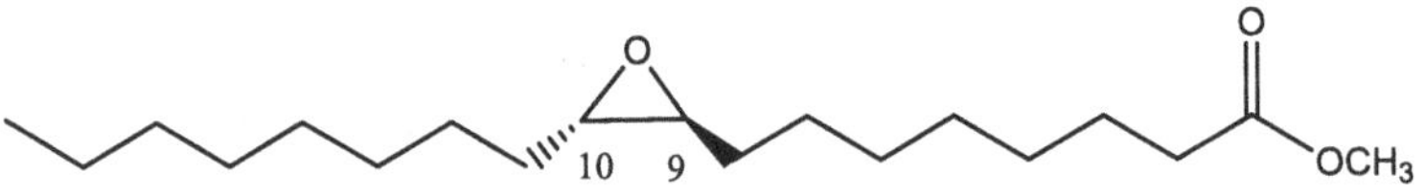

(*E*)-9,10-Epoxystearate

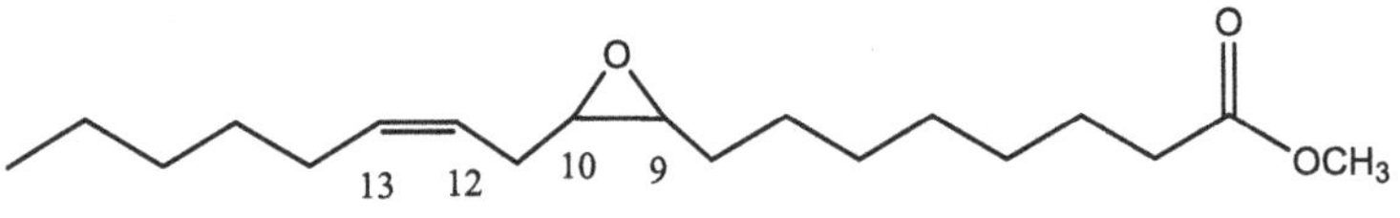

9,10-Epoxy-12-octadecenoate (leukotoxin)

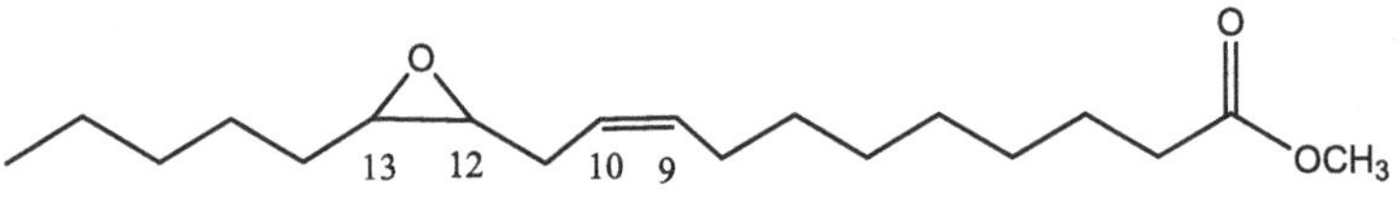

12,13-Epoxy-9-octadecenoate (isoleukotoxin)

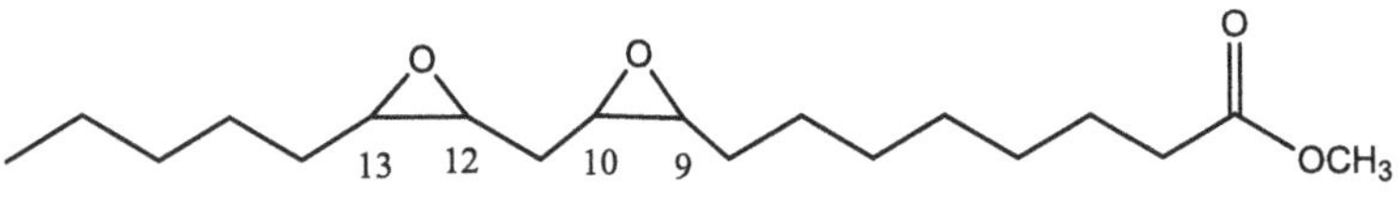

9,10-12,13-Diepoxy-octadecanoate

Fig. 4.5 Structures of epoxides produced from extra virgin olive oil heated at 190 °C (Guillén and Uriarte 2012c)

Table 4.2 Chemical shift and splitting patterns of the ¹H NMR signals of oxidation products produce by heating at frying temperatures or during frying processes (Martínez-Yusta and Guillén 2014a, 2016)

Chemical shift (ppm)	Splitting pattern	Compounds
2.63	Multiplet	(E)-9,10-epoxystearyl groups
2.88,	Multiplet	(Z)-9,10-epoxystearyl groups
3.54–3.59	Multiplet	–CHOH–, secondary alcohols
3.62	Triplet	–CH$_2$OH–, primary alcohols
3.71	Doublet	–CH$_2$OH, 1,2-diglycerides
4.04–4.10	Multiplet	–CHOH, 1,3-diglycerides
9.49	Doublet	–CHO, (E)-2-alkenals
9.52	Doublet	–CHO, (E,E)-2,4-alkadienals
9.60	Doublet	–CHO, (Z,E)-2,4-alkadienals
9.75	Triplet	–CHO, n-alkanals
9.78	Triplet	–CHO, 4-oxo-alkanals

Chapter 5
Use of ^{13}C NMR Spectroscopy for Determination of Lipid Oxidation

^{13}C NMR spectroscopy is not as frequently used as ^{1}H NMR spectroscopy because it is less sensitive than ^{1}H NMR and the intensity of ^{13}C NMR signals are not proportional to the number of ^{13}C atoms. However, it is a very powerful tool to elucidate the molecular structure of a compound, especially when it is used with ^{1}H NMR spectroscopy. A ^{13}C NMR spectrum is typically obtained from the same sample used for ^{1}H NMR. One advantage of the ^{13}C NMR is that the chemical shifts of ^{13}C NMR signals are quite characteristic for specific carbons. For example, the carbonyl carbons of aldehydes and ketones are shown at 180–200 ppm, while carbonyl carbons of acids and esters appear at 160–185 ppm. Signals at 125–150, 115–140, 50–90, 30–65, 16–25, 10–15 ppm can be assigned as carbons of aromatic rings, alkenes, alcohols, amines, aliphatic CH_2, and terminal CH_3, respectively. When it is used for two dimensional (2D) NMR techniques with ^{1}H NMR, such as ^{1}H–^{13}C heteronuclear single-quantum correlation spectroscopy (HSQC), it can give information on which proton is attached to which carbon. The distortionless enhancement by polarization transfer (DEPT), one of the ^{13}C NMR techniques, is used to distinguish CH, CH_2, and CH_3 groups. The attached proton test (APT) technique can also distinguish between carbon atoms with even or odd number of hydrogens. Therefore, the ^{13}C NMR spectroscopy and related techniques have been used for analyses of foods and agricultural products, as well as for lipids and lipid oxidation. In fact, in many studies to identify oxidation products during oil oxidation, ^{13}C NMR has been used along with ^{1}H NMR (Awl et al. 1986; Frankel et al. 1982).

In addition to its supplementary role to ^{1}H NMR, ^{13}C NMR alone can be used for lipid oxidation. For example, Pfeffer et al. (1977) found that the ^{13}C NMR spectroscopy is a good method to distinguish *cis-* and *trans-* isomers of a mono-unsaturated fatty acid. In comparison between ^{13}C NMR spectra of methyl oleate (*cis-*) and methyl elaidate (*trans-*), the *cis*-C9 and C10 signals appeared at 129.8 and 130.0 ppm, while the *trans*-C9 and C10 signals appeared at 130.3 and 130.5 ppm. In addition, chemical shifts of the allylic carbon signals of *cis*-C8, C11 (27.2 and 27.3 ppm) and those of *trans*-C8, C11 (32.5 and 32.6 ppm) were clearly different.

© The Author(s) 2017
H.-S. Hwang, *Advances in NMR Spectroscopy for Lipid Oxidation Assessment*,
SpringerBriefs in Food, Health, and Nutrition, DOI 10.1007/978-3-319-54196-9_5

They also found that the percent *trans*-isomer calculated from the peak intensities of the allylic and olefinic carbon signals agreed well with the actual compositions. The same trend was found with other unsaturated fatty acids (18:2 and 18:3) and therefore, this method can be used to determine the content of *trans*-fats in hydrogenated oils.

Another important application of ^{13}C NMR for lipids is the determination of the fatty acid position in a triacylglycerol. The carbonyl carbons of saturated fatty acid attached to the 1,3-glyceridic carbons appeared at 173.05 ppm in CDCl$_3$ while that attached to the 2-glyceridic carbon appeared at 172.68 ppm (Ng 1983). In another experiment with tripalmitin, triolein, and trilinolein, the chemical shifts of the peaks for the 1,3-positions and 2-position were shown at 173.16 and 172.76 ppm for palmitic acid, 173.13 and 172.72 ppm for oleic acid, and 173.12 and 172.7 ppm for linoleic acid, respectively (Ng 1985). For the fatty acid distribution analysis, the ^{13}C NMR method was found to be notably more convenient than the conventional enzymatic method, which is cumbersome, inaccurate, and cannot be applied to all vegetable oils (Mannina et al. 1999, 2000)

Sacchi et al. (1993) studied lipids extracted from albacore tuna (*Thunnus alalunga*) with ^{13}C NMR. They could conduct simultaneous quantitative determinations of free fatty acids, contents of triacylglycerols and phospholipids, and fatty acid composition and distribution with ^{13}C NMR. The signal of the free fatty acid carbonyl group showed up at a slightly lower field (176.51–177.33 ppm) than those bound to glycerol (172.08–173.83 ppm). A major problem of the ^{13}C NMR method is that the linearity between the signal intensity and the concentration of the component can be distorted, which is caused by a different relaxation rate and/or a different nuclear Overhauser effect (NOE) of carbons considered in the relative calculation. However, after the detailed investigation, they concluded that for accurate quantification of triacylglycerols, the spectra should be acquired in the inverse-gated decoupling mode for complete NOE suppression, while the broad band mode can still be used for rapid quantitative analysis of free fatty acids and semi-quantitative evaluation of triacylglycerols. This ^{13}C NMR method was also used to determine the amount of free fatty acids during industrial canning of tuna (Medina et al. 1994). The ^{13}C NMR method was also used to determine the quantities of mono-, di-, and triacylglycerols in olive oil (Vlahov 1996). Since the ^{13}C NMR technique provides information on the fatty acid composition of oil and the fatty acid distribution, it can be used to ensure the authenticity of edible oils to prevent frauds, such as the addition with esterified oils.

Using the ^{13}C NMR method, Medina et al. (1998) could apply the ^{13}C NMR technique to elucidate the mechanism of lipid oxidation that occurs during the thermal treatment of fish oil. They found that unsaturated fatty acids located at the *sn*-2 position of the glycerol were most prone to oxidative damage during heating of fish oil at 115, 130, and 150 °C. It was also observed that the allylic sites closest to the carbonyl group were the most susceptible to oxidation, followed by those placed near the methyl terminal group. Double bonds located in the middle of the carbon chain showed lower reactivity. The ^{13}C NMR method determining the positional distribution of fatty acids was also applied for the investigation of the oxidative stability of argan oils kept in the dark at 60 °C (Khallouki et al. 2008). The same method was used in the determination of whether an oil sample contained butterfat, lauric oils, partially hydrogenated fat, linoleic acid, or linolenic acid (Gunstone 1993).

Hidalgo et al. (2002) attempted to predict the stability of the non-polar portion of vegetable oil using ^{13}C NMR signals. They fractioned 66 vegetable oils using column chromatography, obtained ^{13}NMR spectra, analyzed fatty acids and triacylglycerol composition, and determined oil stability by the Rancimat. By using some selected signals in ^{13}C NMR spectrum, they found that the predicted stabilities calculated by the NMR data were always better than those obtained by using chemical determinations including phenol content, tocopherol content, and fatty acid composition palmitic acid, since ^{13}C NMR considered many more variables than those determined by chemical analysis.

One useful ^{13}C NMR technique is the distortionless enhancement by polarization transfer (DEPT), with which one can distinguish quaternary (C), methine (CH), methylene (CH$_2$), and methyl (CH$_3$) carbons, unlike the normal ^{13}C NMR. Since modern analytical techniques can verify complicated molecular structures of oxidation products, deeper understanding of lipid oxidation can be achieved using these techniques. For example, the DEPT technology was used to study the molecular structure of polymers formed during the oxidation of soybean oil, and it was found that the generally accepted Diels-Alder reaction did not occur (Doll and Hwang 2013; Hwang et al. 2013a). The Diels-Alder reaction mechanism has been believed to be one of the major mechanisms to produce dimers, oligomers, and polymers during the oxidation of edible oil since the 1930s. In the DEPT135 spectrum, CH and CH$_3$ signals are shown in a phase opposite to CH$_2$ signals. Quaternary carbons (C) do not appear in the DEPT 135. Therefore, signals appeared in the normal ^{13}C NMR spectrum but not in the DEPT 135 spectrum are identified as quaternary carbons (C). Figure 5.1 shows the mechanism of the Diels-Alder reaction to produce a six-membered ring compound during oil oxidation. The double bonds in the oil molecule undergo the rearrangement to produce a conjugated diene, which, in turn, reacts with another double bond to produce the six-membered ring. If the Diels-Alder reaction occurred, the new methine carbons (CH, carbons a–d in Fig. 5.1) should appear in the region of 25–60 ppm in the DEPT spectrum as negative peaks. However, no new methine peaks were observed with soybean oil oxidized at 180 °C for 24 h, although the oxidized oil had 36% polymers. In another experiment, a fraction containing 92% polymers was separated using a column chromatography and the DEPT spectrum was taken for this fraction. However, this fraction did not show any new methine signals corresponding to the Diels-Alder products either (Fig. 5.1).

Even with methyl conjugated linoleate, which was supposed to undergo the Diels-Alder reaction more easily than soybean oil containing non-conjugated double bonds, no product indicating this reaction was observed in the DEPT spectrum. On the other hand, methyl oleate and triolein that do not contain a diene and cannot undergo the Diels-Alder reaction produced significant amounts of polymers (27.2% and 63.0%, respectively) after heating at 180 °C for 24 h. Therefore, the research group concluded that the Diels-Alder reaction was not the major reaction mechanism to produce polymers in the oxidation of edible oil.

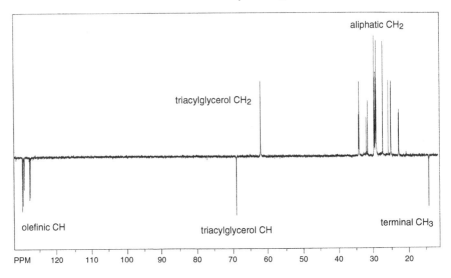

Fig. 5.1 The proposed Diels-Alder reaction and the DEPT spectrum of a fraction of oxidized soybean oil containing 92% polymers (Hwang and Bakota 2015; Hwang et al. 2013a)

The same Diels-Alder reaction mechanism was believed for a long time to be one of the major reactions for the polymerization of anaerobically heated oil. However, no Diels-Alder reaction products were detected while heating soybean oil under nitrogen when examined with the DEPT (Arca et al. 2012). Therefore, the radical-initiated polymerization reactions are considered to be more possible reactions than the Diels-Alder reaction for polymerization of soybean oil and other edible oils. Patrikios and Mavromoustakos (2014) found that the dimers and polymers had an ether linkage between fatty acids in virgin olive oil heated at 130 °C for 24 h using ^{1}H and ^{13}C NMR spectroscopies, gas chromatography-mass spectroscopy (GC-MS), and two-dimensional (2D) NMR (HSQC) spectroscopy for the structural characterization.

Chapter 6
³¹P NMR Spectroscopy for Assessment of Lipid Oxidation

The ³¹P NMR spectroscopy is known to be a better analytical method than time-consuming HPLC and TLC techniques in analyses of phospholipids in oil such as phosphatidylcholine, phosphatidylethanolamine, phosphatidylglycerol, and phosphatidic acid (Pearce and Komoroski 1993). Hatzakis et al. (2008) used the ³¹P NMR spectroscopy to determine phospholipids extracted from extra virgin olive oil using a mixture of ethanol:water (2:1, v/v). A satisfactory sensitivity was observed to have the detection limit of 0.25–1.24 µmol/mL.

The ³¹P NMR spectroscopy is very useful in case where strong signal overlap and dynamic range problems in ¹H NMR spectra and/or long relaxation time of ¹³C NMR become a problem (Dais and Hatzakis 2013). The ³¹P NMR spectroscopy can also be used to determine the fatty acid distribution and to assess the oxidation of vegetable oils. The major drawback of ³¹P NMR for this application compared to ¹H and ¹³C NMR, is that the experimental procedure involves an additional step, the derivatization of functional groups (−OH and −COOH) in the oil sample with a phosphorus reagent, 2-chloro-4,4,5,5-tetramethyldioxaphospholane. Although this reaction occurs in the NMR tube and is completed in less than 15 min after the addition of the phosphorus reagent, the additional experimental step is its major disadvantage. Figure 6.1 shows the derivatization of functional groups with the phosphorus reagent producing a phosphorus-containing five-membered ring, which resonates in the range 100–200 ppm.

The much higher sensitivity over ¹³C NMR, the large range of chemical shifts relative to ¹H and ¹³C NMR, and the resulting good separation of signals are advantages of this technique. Therefore, one of the major uses of this technique is to distinguish structurally similar compounds such as mono-, di-, and triacylglycerols. For example, the excellent resolution between the ³¹P NMR chemical shifts of monoacylglycerols and diacylglycerols allows the reliable quantification of these molecules in olive oil (Dais and Hatzakis 2013). Spyros et al. (2004) could quantitatively determine 1,2- and 1,3-diacylglycerol isomers of extra virgin olive oil samples with ³¹P NMR methodology as a function of storage time and storage conditions.

© The Author(s) 2017

H.-S. Hwang, *Advances in NMR Spectroscopy for Lipid Oxidation Assessment*,
SpringerBriefs in Food, Health, and Nutrition,
DOI 10.1007/978-3-319-54196-9_6

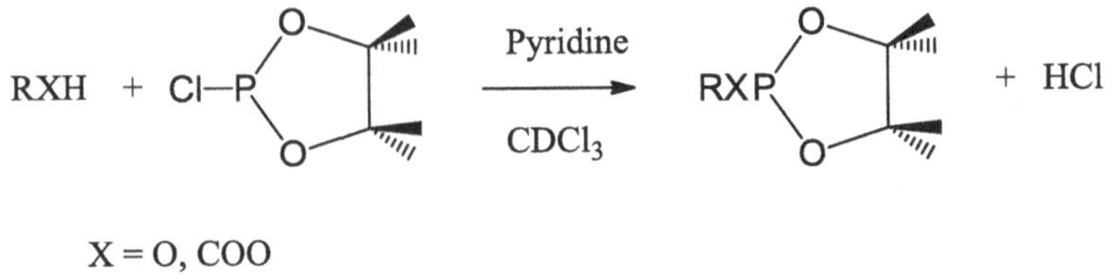

X = O, COO

Fig. 6.1 Reaction of hydroxyl and carboxylic acid groups of oil sample with 2-chloro-4,4, 5,5-tetramethyldioxaphospholane

Spyros and Dais (2000) provided the database for specific chemical shifts of the hydroxyl group of mono-and diacylglycerols and carboxylic acid groups of fatty acids from the study with model compounds including 1,3-diolein, 1,2-diolein, 1-monoolein, and 2-monoolein, and many free fatty acids. Using this database, they could determine the quantities of 1-monoacylglycerol, 1,3-diacylglycerol and 1,2-diglycerol in several olive oil samples from different regions of Greece and different years of production. Dayrit et al. (2011) used the ^{31}P NMR method to determine the rate of hydrolysis of virgin coconut oil at 30 and 80 °C. Lucas-Torres et al. (2014) found that the 1,3-diacylglycerols have greater thermal stability than 1,2-diacylglycerols when they monitored the connection changes of 1,2- and 1,3-diacylglycerols using the ^{31}P NMR spectroscopy during heating of olive oil.

Some other applications of ^{13}P NMR spectroscopy include the quantification of moisture in olive oils (Hatzakis and Dais 2008) and the determination of phenolic compounds in the polar fraction of virgin olive oil (Christophoridou and Dais 2006). 2D ^{31}P NMR, such as homonuclear ^{31}P–^{31}P and heteronuclear ^{1}H–^{31}P spectroscopies, were also utilized for the analysis of lipid.

Chapter 7
Conclusions and Future Prospects

Lipid oxidation involves many complicated reactions and therefore, it is still very difficult to understand these reactions, analyze numerous oxidation products, and accurately determine the level of oxidation. Even though there are a variety of analytical methods developed for the determination of lipid oxidation, many of them are old methods that depend on chemical reactions with a reagent, require long time and extensive labor, cannot be automated, measure only one kind of oxidation product, determine the concentration of an oxidation product that reaches a peak in a short time, and/or have inconsistencies due to multiple variations in the procedure. Extremely reliable modern analytical instruments, which can handle many samples at a time by automation, concomitantly detect many oxidation products, and be used for a long period of oxidation process, should be utilized to overcome the problems of current standard analytical methods.

According to the literatures, the ^{1}H NMR method for the assessment of lipid oxidation meets many of these needs providing; (1) shorter time and convenience, (2) less variables for human errors and exceptionally high reproducibility and repeatability (simply an approximate amount of the sample is dissolved in an NMR solvent), and (3) high reliability. The ^{1}H NMR is also a very reliable tool to identify numerous oxidation products and, therefore, the ^{1}H NMR techniques are highly recommended for deeper understanding of oxidation mechanisms. ^{13}C NMR spectroscopy is a very powerful tool for the elucidation of molecular structures and the determination of fatty acid composition and distribution. ^{31}P NMR spectroscopy can be used as a reliable method to determine fatty acid distribution and also to assess oxidation of vegetable oils.

The NMR instrument is in general very expensive and many small laboratories cannot afford this instrument. This seems to be one of the major reasons that the NMR method is not widely used for the assessment of lipid oxidation. Another hurdle for the utilization of the NMR method is that many scientists are not very familiar with the NMR instrument and think that it is complicated and difficult to use. However, it should be noted that some instruments, such as the instrument

© The Author(s) 2017
H.-S. Hwang, *Advances in NMR Spectroscopy for Lipid Oxidation Assessment*,
SpringerBriefs in Food, Health, and Nutrition,
DOI 10.1007/978-3-319-54196-9_7

measuring solid fat contents (SFC), are small benchtop NMR instruments being widely used. Low-resonance NMR instruments, such as a 90 MHz NMR, also provided reliable data for lipid oxidation. Therefore, while the high-resolution NMR instruments can be used for those people who already have one in their labs, new NMR instruments specifically designed to measure lipid oxidation can be developed. In fact, recent efforts to provide more convenient, user-friendly NMR methods were made. For example, the automated analysis of vegetable oils by low-field NMR to obtain the fatty acid composition was reported (Castejón et al. 2016). Furthermore, compact low-field NMR instruments were developed, which successfully provided decent ^{1}H NMR, ^{13}C NMR, ^{9}F NMR, ^{31}P NMR, and 2D NMR spectra (Blümich 2016). One of the problems of 2D NMR is the long experimental time, which is not suitable for high-throughput applications such as real-time reaction monitoring or rapid screening. To overcome this problem, a benchtop NMR spectrometer that can perform the 2D NMR analysis in a short time has been developed (Gouilleux et al. 2016). Therefore, it is believed that the NMR method will be widely used to determine lipid oxidation and will be thought of as a standard method in the near future.

Furthermore, the current criterion for the safety level of oxidized oil is the amount of polar compounds in oil, while it was reported that the amount of polar compounds could not be representative of the safety level. It was reported that oils with the same amount of polar compounds had different compositions of oxidation products, such as genotoxic and cytotoxic 4-hydroxy-alkenals (Guillén and Uriarte 2012c). The old criterion for the safety level of edible oil should be re-considered. Since the NMR method can be used not only to monitor the oxidation of lipids, but also to identify and quantify a specific oxidation product, it can be a better method to determine the safety level of oil.

References

Abou-Gharbia HA, Shahidi F, Shehata AAY, Youssef MM (1996) Oxidative stability of extracted sesame oil from raw and processed seeds. J Food Lipids 3:59–72. doi:10.1111/j.1745-4522.1996.tb00054.x

Abou-Gharbia HA, Shahidi F, Adel A, Shehata Y, Youssef MM (1997) Effects of processing on oxidative stability of sesame oil extracted from intact and dehulled seeds. J Am Oil Chem Soc 74(3):215–221. doi:10.1007/s11746-997-0126-9

Alonso-Salces RM, Héberger K, Holland MV, Moreno-Rojas JM, Mariani C, Bellan G, Reniero F, Guillou C (2010) Multivariate analysis of NMR fingerprint of the unsaponifiable fraction of virgin olive oils for authentication purposes. Food Chem 118(4):956–965. doi:10.1016/j.foodchem.2008.09.061

Antolovich M, Prenzler PD, Patsalides E, McDonald S, Robards K (2002) Methods for testing antioxidant activity. Analyst 127(1):183–198. doi:10.1039/b009171p

AOAC (1996) Polymerized triglycerides in oils and fats. Gel-permeation liquid chromatographic method. vol Test Method:AOAC 993.25-1996

Arca M, Sharma B, Price NJ, Perez J, Doll K (2012) Evidence contrary to the accepted Diels–Alder mechanism in the thermal modification of vegetable oil. J Am Oil Chem Soc 89(6):987–994. doi:10.1007/s11746-011-2002-x

Augustin MA, Berry SK (1983) Efficacy of the antioxidants BHA and BHT in palm olein during heating and frying. J Am Oil Chem Soc 60(8):1520–1523. doi:10.1007/bf02666575

Awl RA, Neff WE, Frankel EN, Plattner RD, Weisleder D (1986) Cyclic fatty esters: hydroperoxides from photosensitized oxidation of methyl 9-(6-propyl-3-cyclohexenyl)-(Z)8-nonenoate. Chem Phys Lipids 39(1-2):1–17. doi:10.1016/0009-3084(86)90095-2

Bakota EL, Winkler-Moser JK, Palmquist DE (2012) Solid fat content as a substitute for total polar compound analysis in edible oils. J Am Oil Chem Soc 89(12):2135–2142. doi:10.1007/s11746-012-2121-z

Barriuso B, Astiasarán I, Ansorena D (2013) A review of analytical methods measuring lipid oxidation status in foods: a challenging task. Eur Food Res Technol 236(1):1–15. doi:10.1007/s00217-012-1866-9

Barthel G, Grosch W (1974) Peroxide value determination—comparison of some methods. J Am Oil Chem Soc 51(12):540–544. doi:10.1007/bf02636025

Bathista ALBS, Silva EOD, Tavares MIB, Prado RJ (2012) Solid-state NMR to evaluate the molecular changes in the mango starch after 8 years of storage. J Appl Polym Sci 126(S1):E123–E126. doi:10.1002/app.36703

Blumenthal MM, Trout JR, Chang SS (1976) Correlation of gas chromatographic profiles and organoleptic scores of different fats and oils after simulated deep fat frying. J Am Oil Chem Soc 53(7):496–501. doi:10.1007/bf02636822

© The Author(s) 2017

51

H.-S. Hwang, *Advances in NMR Spectroscopy for Lipid Oxidation Assessment*, SpringerBriefs in Food, Health, and Nutrition, DOI 10.1007/978-3-319-54196-9

Blümich B (2016) Introduction to compact NMR: a review of methods. Trends Anal Chem 83:2–11. doi:10.1016/j.trac.2015.12.012

Bolland JL, Koch HP (1945) The course of autoxidation reactions in polyisoprenes and allied compounds. Part IX. The primary thermal oxidation product of ethyl linoleate. J Chem Soc 1945:445–447. doi:10.1039/jr9450000445

Castejón D, Mateos-Aparicio I, Molero MD, Cambero MI, Herrera A (2014) Evaluation and optimization of the analysis of fatty acid types in edible oils by ^{1}H-NMR. Food Anal Methods 7(6):1285–1297. doi:10.1007/s12161-013-9747-9

Castejón D, Fricke P, Cambero MI, Herrera A (2016) Automatic ^{1}H-NMR screening of fatty acid composition in edible oils. Nutrients 8(2). doi:10.3390/nu8020093

Cazor A, Deborde C, Moing A, Rolin D, This H (2006) Sucrose, glucose, and fructose extraction in aqueous carrot root extracts prepared at different temperatures by means of direct NMR measurements. J Agric Food Chem 54(13):4681–4686. doi:10.1021/jf060144i

Chakraborty K, Joseph D (2016) Changes in the quality of refined fish oil in an accelerated storage study. J Aquat Food Prod Technol 25(7):1–16. doi:10.1080/10498850.2015.1036482

Chan HS, Levett G (1977) Autoxidation of methyl linoleate. Separation and analysis of isomeric mixtures of methyl linoleate hydroperoxides and methyl hydroxylinoleates. Lipids 12(1):99–104. doi:10.1007/bf02532979

Chatgilialoglu C, Ferreri C (2010) Biomimetic chemistry: radical reactions in vesicle suspensions. In: Amitava M (ed) Biomimetics learning from nature. InTech, Rijeka. doi:10.5772/8774

Choe E, Min DB (2006) Mechanisms and factors for edible oil oxidation. Compr Rev Food Sci Food Saf 5(4):169–186. doi:10.1111/j.1541-4337.2006.00009.x

Christophoridou S, Dais P (2006) Novel approach to the detection and quantification of phenolic compounds in olive oil based on ^{31}P nuclear magnetic resonance spectroscopy. J Agric Food Chem 54(3):656–664. doi:10.1021/jf058138u

Claxson AWD, Hawkes GE, Richardson DP, Naughton DP, Haywood RM, Chander CL, Atherton M, Lynch EJ, Grootveld MC (1994) Generation of lipid peroxidation products in culinary oils and fats during episodes of thermal stressing: a high field ^{1}H NMR study. FEBS Lett 355(1):81–90. doi:10.1016/0014-5793(94)01147-8

Colzato M, Scramin JA, Forato LA, Colnago LA, Assis OBG (2011) ^{1}H NMR investigation of oil oxidation in macadamia nuts coated with zein-based films. J Food Process Pres 35(6):790–796. doi:10.1111/j.1745-4549.2011.00530.x

Correia AC, Dubreucq E, Ferreira-Dias S, Lecomte J (2015) Rapid quantification of polar compounds in thermo-oxidized oils by HPTLC-densitometry. Eur J Lipid Sci Technol 117(3):311–319. doi:10.1002/ejlt.201400230

Dais P, Hatzakis E (2013) Quality assessment and authentication of virgin olive oil by NMR spectroscopy: a critical review. Anal Chim Acta 765:1–27. doi:10.1016/j.aca.2012.12.003

Dayrit FM, Dimzon IKD, Valde MF, Santos JER, Garrovillas MJM, Villarino BJ (2011) Quality characteristics of virgin coconut oil: comparisons with refined coconut oil. Pure Appl Chem 83(9):1789–1799. doi:10.1351/pac-con-11-04-01

Dobarganes MC (2009) Formation of volatiles and short-chain bound compounds. AOCS. http://lipidlibrary.aocs.org/OilsFats/content.cfm?ItemNumber=39213. Accessed 10 Jan 2017

Dobarganes MC, Velasco J (2002) Analysis of lipid hydroperoxides. Eur J Lipid Sci Technol 104(7):420–428. doi:10.1002/1438-9312(200207)104:7<420::aid-ejlt420>3.0.co;2-n

Doll KM, Hwang H-S (2013) Thermal modification of vegetable oils. Lipid Technol 25(4):83–85. doi:10.1002/lite.201300269

Falch E, Anthonsen HW, Axelson DE, Aursand M (2004) Correlation between ^{1}H NMR and traditional methods for determining lipid oxidation of ethyl docosahexaenoate. J Am Oil Chem Soc 81(12):1105–1110. doi:10.1007/s11746-004-1025-1

Farhoosh R, Moosavi SMR (2009) Evaluating the performance of peroxide and conjugated diene values in monitoring quality of used frying oils. J Agric Sci Technol 11(2):173–179

Farhoosh R, Tavassoli-Kafrani MH (2011) Simultaneous monitoring of the conventional qualitative indicators during frying of sunflower oil. Food Chem 125(1):209–213. doi:10.1016/j.foodchem.2010.08.064

Fernández J, Pérez-Álvarez JA, Fernández-López JA (1997) Thiobarbituric acid test for monitoring lipid oxidation in meat. Food Chem 59(3):345–353. doi:10.1016/S0308-8146(96)00114-8

Fhaner M, Hwang HS, Winkler-Moser JK, Bakota EL, Liu SX (2016) Protection of fish oil from oxidation with sesamol. Eur J Lipid Sci Technol 118(6):885–897. doi:10.1002/ejlt.201500185

Frankel EN (2012a) Methods to determine extent of oxidation. In: Decker EA (ed) Lipid oxidation. Woodhead Publishing, Sawston, pp 99–127. doi:10.1533/9780857097927.99

Frankel EN (2012b) Introduction. In: Decker EA (ed) Lipid oxidation. Woodhead Publishing, Sawston, pp 1–14. doi:10.1533/9780857097927.1

Frankel EN, Neff WE, Selke E, Weisleder D (1982) Photosensitized oxidation of methyl linoleate: secondary and volatile thermal decomposition products. Lipids 17(1):11–18. doi:10.1007/bf02535116

Gertz C (2000) Chemical and physical parameters as quality indicators of used frying fats. Eur J Lipid Sci Technol 102(8-9):566–572. doi:10.1002/1438-9312(200009)102:8/9<566::aid-ejlt 566>3.0.co;2-b

Goicoechea E, Guillen MD (2010) Analysis of hydroperoxides, aldehydes and epoxides by ^{1}H nuclear magnetic resonance in sunflower oil oxidized at 70 and 100 °C. J Agric Food Chem 58(10):6234–6245. doi:10.1021/jf1005337

Gouilleux B, Charrier B, Akoka S, Felpin FX, Rodriguez-Zubiri M, Giraudeau P (2016) Ultrafast 2D NMR on a benchtop spectrometer: applications and perspectives. Trends Anal Chem 83:65–75. doi:10.1016/j.trac.2016.01.014

Gray JI (1978) Measurement of lipid oxidation: a review. J Am Oil Chem Soc 55(6):539–546. doi:10.1007/bf02668066

Guillen MD, Goicoechea E (2009) Oxidation of corn oil at room temperature: primary and secondary oxidation products and determination of their concentration in the oil liquid matrix from ^{1}H nuclear magnetic resonance data. Food Chem 116(1):183–192. doi:10.1016/j.foodchem.2009.02.029

Guillén MD, Ruiz A (2001) High resolution ^{1}H nuclear magnetic resonance in the study of edible oils and fats. Trends Food Sci Tech 12(9):328–338. doi:10.1016/S0924-2244(01)00101-7

Guillén MD, Ruiz A (2003a) ^{1}H nuclear magnetic resonance as a fast tool for determining the composition of acyl chains in acylglycerol mixtures. Eur J Lipid Sci Technol 105(9):502–507. doi:10.1002/ejlt.200300799

Guillén MD, Ruiz A (2003b) Rapid simultaneous determination by proton NMR of unsaturation and composition of acyl groups in vegetable oils. Eur J Lipid Sci Technol 105(11):688–696. doi:10.1002/ejlt.200300866

Guillén MD, Ruiz A (2004) Formation of hydroperoxy- and hydroxyalkenals during thermal oxidative degradation of sesame oil monitored by proton NMR. Eur J Lipid Sci Technol 106(10):680–687. doi:10.1002/ejlt.200401026

Guillén MD, Ruiz A (2005a) Monitoring the oxidation of unsaturated oils and formation of oxygenated aldehydes by proton NMR. Eur J Lipid Sci Technol 107(1):36–47. doi:10.1002/ejlt.200401056

Guillén MD, Ruiz A (2005b) Oxidation process of oils with high content of linoleic acyl groups and formation of toxic hydroperoxy- and hydroxyalkenals. A study by ^{1}H nuclear magnetic resonance. J Sci Food Agric 85(14):2413–2420. doi:10.1002/jsfa.2273

Guillén MD, Ruiz A (2005c) Study by proton nuclear magnetic resonance of the thermal oxidation of oils rich in oleic acyl groups. J Am Oil Chem Soc 82(5):349–355. doi:10.1007/s11746-005-1077-2

Guillén MD, Uriarte PS (2009) Contribution to further understanding of the evolution of sunflower oil submitted to frying temperature in a domestic fryer: study by ^{1}H nuclear magnetic resonance. J Agric Food Chem 57(17):7790–7799. doi:10.1021/jf900510k

Guillén MD, Uriarte PS (2012a) Monitoring by ^{1}H nuclear magnetic resonance of the changes in the composition of virgin linseed oil heated at frying temperature. Comparison with the evolution of other edible oils. Food Control 28(1):59–68. doi:10.1016/j.foodcont.2012.04.024

Guillén MD, Uriarte PS (2012b) Simultaneous control of the evolution of the percentage in weight of polar compounds, iodine value, acyl groups proportions and aldehydes concentrations in sunflower oil submitted to frying temperature in an industrial fryer. Food Control 24(1-2): 50–56. doi:10.1016/j.foodcont.2011.09.002

Guillén MD, Uriarte PS (2012c) Study by ¹H NMR spectroscopy of the evolution of extra virgin olive oil composition submitted to frying temperature in an industrial fryer for a prolonged period of time. Food Chem 134(1):162–172. doi:10.1016/j.foodchem.2012.02.083

Guillén MD, Uriarte PS (2013) Relationships between the evolution of the percentage in weight of polar compounds and that of the molar percentage of acyl groups of edible oils submitted to frying temperature. Food Chem 138(2–3):1351–1354. doi:10.1016/j.foodchem.2012.10.108

Gunstone FD (1993) Information on the composition of fats from their high-resolution ¹³C nuclear magnetic resonance spectra. J Am Oil Chem Soc 70(4):361–366. doi:10.1007/bf02552707

Gunstone FD, Hilditch TP (1945) The union of gaseous oxygen with methyl oleate, linoleate, and linolenate. J Chem Soc 1945:836–841. doi:10.1039/jr9450000836

Hara S, Totani Y (1988) A highly sensitive method for the micro-determination of lipid hydroperoxides by potentiometry. J Am Oil Chem Soc 65(12):1948–1950. doi:10.1007/bf02546014

Hatzakis E, Dais P (2008) Determination of water content in olive oil by ³¹P NMR spectroscopy. J Agric Food Chem 56(6):1866–1872. doi:10.1021/jf073227n

Hatzakis E, Koidis A, Boskou D, Dais P (2008) Determination of phospholipids in olive oil by ³¹P NMR spectroscopy. J Agric Food Chem 56(15):6232–6240. doi:10.1021/jf800690t

Hein M, Henning H, Isengard HD (1998) Determination of total polar parts with new methods for the quality survey of frying fats and oils. Talanta 47(2):447–454. doi:10.1016/s0039-9140(98)00148-9

Hein M, Isengard HD (1997) The use of high performance liquid chromatography for the quality survey of frying fats and oils. Chromatographia 45:373–377. doi:10.1007/bf02505587

Herrero AM (2008) Raman spectroscopy a promising technique for quality assessment of meat and fish: a review. Food Chem 107(4):1642–1651. doi:10.1016/j.foodchem.2007.10.014

Hicks M, Gebicki JM (1979) A spectrophotometric method for the determination of lipid hydroperoxides. Anal Biochem 99(2):249–253. doi:10.1016/S0003-2697(79)80003-2

Hidalgo FJ, Gómez G, Navarro JL, Zamora R (2002) Oil stability prediction by high-resolution ¹³C nuclear magnetic resonance spectroscopy. J Agric Food Chem 50(21):5825–5831. doi:10.1021/jf0256539

Holman RT, Elmer OC (1947) The rates of oxidation of unsaturated fatty acids and esters. J Am Oil Chem Soc 24(4):127–129. doi:10.1007/bf02643258

Hwang H-S (2015) NMR spectroscopy for assessing lipid oxidation. Lipid Technol 27(8): 187–189. doi:10.1002/lite.201500037

Hwang H-S, Bakota E (2015) NMR spectroscopy for evaluation of lipid oxidation. In: Rahman AU, Choudhary MI (eds) Application of NMR spectroscopy, Bentham eBooks, vol 4. Oak Park, IL, pp 62–95. doi:10.2174/9781681081434116040l

Hwang H-S, Doll KM, Winkler-Moser JK, Vermillion K, Liu SX (2013a) No evidence found for Diels–Alder reaction products in soybean oil oxidized at the frying temperature by NMR study. J Am Oil Chem Soc 90(6):825–834. doi:10.1007/s11746-013-2229-9

Hwang H-S, Winkler-Moser JK (2014) Food additives reducing volatility of antioxidants at frying temperature. J Am Oil Chem Soc 91(10):1745–1761. doi:10.1007/s11746-014-2525-z

Hwang H-S, Winkler-Moser JK, Bakota EL, Berhow MA, Liu SX (2013b) Antioxidant activity of sesamol in soybean oil under frying conditions. J Am Oil Chem Soc 90(5):659–666. doi:10.1007/s11746-013-2204-5

Hwang H-S, Winkler-Moser JK, Liu SX (2012) Structural effect of lignans and sesamol on polymerization of soybean oil at frying temperature. J Am Oil Chem Soc 89(6):1067–1076. doi:10.1007/s11746-011-1994-6

Hwang H-S, Winkler-Moser JK, Liu SX (2017) Reliability of ¹H NMR analysis for assessment of lipid oxidation at frying temperatures. J Am Oil Chem Soc:1–14. doi:10.1007/s11746-016-2945-z

Hwang H-S, Winkler-Moser JK, Vermillion K, Liu SX (2014) Enhancing antioxidant activity of sesamol at frying temperature by addition of additives through reducing volatility. J Food Sci 79(11):C2164–C2173. doi:10.1111/1750-3841.12653

Iulianelli GCV, Tavares MIB (2016) Application of solid-state NMR spectroscopy to evaluate cassava genotypes. J Food Compos Anal 48:88–94. doi:10.1016/j.jfca.2016.02.009

IUPAC (1992) Standard methods for the analysis of oils, fats and derivatives, IUPAC standard method 2.501: determination of peroxide value. Pergamon, International Union of Pure and Applied Chemistry

Jiang Z-Y, Woollard ACS, Wolff SP (1991) Lipid hydroperoxide measurement by oxidation of Fe^{2+} in the presence of xylenol orange. Comparison with the TBA assay and an iodometric method. Lipids 26(10):853–856. doi:10.1007/bf02536169

Kanner J, Rosenthal I (1992) An assessment of lipid oxidation in foods. Pure Appl Chem 64(12):1959–1964. doi:10.1351/pac199264121959

Karoui R, Blecker C (2011) Fluorescence spectroscopy measurement for quality assessment of food systems – a review. Food Bioproc Tech 4(3):364–386. doi:10.1007/s11947-010-0370-0

Kaufmann A, Ryser B, Suter B (2001) HPLC with evaporative light scattering detection for the determination of polar compounds in used frying oils. Eur Food Res Technol 213(4-5):372–376. doi:10.1007/s002170100373

Khallouki F, Mannina L, Viel S, Owen RW (2008) Thermal stability and long-chain fatty acid positional distribution on glycerol of argan oil. Food Chem 110(1):57–61. doi:10.1016/j.foodchem.2008.01.055

Khatoon S, Krishna AGG (1998) Assessment of oxidation in heated safflower oil by physical, chemical and spectroscopic methods. J Food Lipids 5(4):247–267

Knothe G, Kenar JA (2004) Determination of the fatty acid profile by ^{1}H-NMR spectroscopy. Eur J Lipid Sci Technol 106(2):88–96. doi:10.1002/ejlt.200300880

Kong F, Singh RP (2011) Advances in instrumental methods to determine food quality deterioration. In: Kilcast D, Subramaniam P (eds) Food and beverage stability and shelf life. Woodhead Publishing, Sawston, pp 381–404. doi:10.1016/b978-1-84569-701-3.50012-8

Laguerre M, Lecomte J, Villeneuve P (2007) Evaluation of the ability of antioxidants to counteract lipid oxidation: existing methods, new trends and challenges. Prog Lipid Res 46(5):244–282. doi:10.1016/j.plipres.2007.05.002

Lea CH (1952) Methods for determining peroxide in lipids. J Sci Food Agric 3(12):586–594. doi:10.1002/jsfa.2740031205

List GR, Evans CD, Kwolek WF, Warner K, Boundy BK, Cowan JC (1974) Oxidation and quality of soybean oil: a preliminary study of the anisidine test. J Am Oil Chem Soc 51(2):17–21. doi:10.1007/bf02545207

Lucas-Torres C, Pérez Á, Cabañas B, Moreno A (2014) Study by ^{31}P NMR spectroscopy of the triacylglycerol degradation processes in olive oil with different heat-transfer mechanisms. Food Chem 165:21–28. doi:10.1016/j.foodchem.2014.05.092

Mannina L, Luchinat C, Emanuele MC, Segre A (1999) Acyl positional distribution of glycerol tri-esters in vegetable oils: a ^{13}C NMR study. Chem Phys Lipids 103(1–2):47–55. doi:10.1016/S0009-3084(99)00092-4

Mannina L, Luchinat C, Patumi M, Emanuele MC, Rossi E, Segre A (2000) Concentration dependence of ^{13}C NMR spectra of triglycerides: implications for the NMR analysis of olive oils. Magn Reson Chem 38(10):886–890. doi:10.1002/1097-458X(200010)38:10<886::AID-MRC738>3.0.CO;2-J

Mannina L, Sobolev AP, Viel S (2012) Liquid state ^{1}H high field NMR in food analysis. Prog Nucl Magn Reson Spectrosc 66:1–39. doi:10.1016/j.pnmrs.2012.02.001

Marcone MF, Wang S, Albabish W, Nie S, Somnarain D, Hill A (2013) Diverse food-based applications of nuclear magnetic resonance (NMR) technology. Food Res Int 51(2):729–747. doi:10.1016/j.foodres.2012.12.046

Mariette F (2009) Investigations of food colloids by NMR and MRI. Curr Opin Colloid Interface Sci 14(3):203–211. doi:10.1016/j.cocis.2008.10.006

Marsili RT (2000) Shelf-life prediction of processed milk by solid-phase microextraction, mass spectrometry, and multivariate analysis. J Agric Food Chem 48(8):3470–3475. doi:10.1021/jf000177c

Martínez-Yusta A, Guillén MD (2014a) Deep-frying food in extra virgin olive oil: a study by ^{1}H nuclear magnetic resonance of the influence of food nature on the evolving composition of the frying medium. Food Chem 150:429–437. doi:10.1016/j.foodchem.2013.11.015

Martínez-Yusta A, Guillén MD (2014b) Deep-frying. A study of the influence of the frying medium and the food nature, on the lipidic composition of the fried food, using ^{1}H nuclear magnetic resonance. Food Res Int 62:998–1007. doi:10.1016/j.foodres.2014.05.015

Martínez-Yusta A, Guillén MD (2016) Monitoring compositional changes in sunflower oil-derived deep-frying media by ^{1}H nuclear magnetic resonance. Eur J Lipid Sci Technol 118(7):984–996. doi:10.1002/ejlt.201500270

Medina I, Sacchi R, Aubourg S (1994) ^{13}C nuclear magnetic resonance monitoring of free fatty acid release after fish thermal processing. J Am Oil Chem Soc 71(5):479–482. doi:10.1007/bf02540657

Medina I, Sacchi R, Giudicianni I, Aubourg S (1998) Oxidation in fish lipids during thermal stress as studied by ^{13}C nuclear magnetic resonance spectroscopy. J Am Oil Chem Soc 75(2):147–154. doi:10.1007/s11746-998-0026-7

Milinsk MC, Matsushita M, Visentainer JV, CCd O, Souza NE (2008) Comparative analysis of eight esterification methods in the quantitative determination of vegetable oil fatty acid methyl esters (FAME). J Braz Chem Soc 19:1475–1483. doi:10.1590/S0103-50532008000800006

Navas MJ, Jiménez AM (1996) Review of chemiluminescent methods in food analysis. Food Chem 55(1):7–15. doi:10.1016/0308-8146(95)00058-5

Neff WE, Frankel EN, Selke E, Weisleder D (1983) Photosensitized oxidation of methyl linoleate monohydroperoxides: hydroperoxy cyclic peroxides, dihydroperoxides, keto esters and volatile thermal decomposition products. Lipids 18(12):868–876. doi:10.1007/bf02534564

Ng S (1983) High resolution [13]C n.m.r. spectra of the carbonyl carbons of the triglycerides of palm oil. J Chem Soc Chem Comm 1983(4):179–180. doi:10.1039/c39830000179

Ng S (1985) Analysis of positional distribution of fatty acids in palm oil by ^{13}C NMR spectroscopy. Lipids 20(11):778–782. doi:10.1007/bf02534402

Nieva-Echevarría B, Goicoechea E, Manzanos MJ, Guillén MD (2016) The influence of frying technique, cooking oil and fish species on the changes occurring in fish lipids and oil during shallow-frying, studied by ^{1}H NMR. Food Res Int 84:150–159. doi:10.1016/j.foodres.2016.03.033

Patrikios IS, Mavromoustakos TM (2014) Monounsaturated fatty acid ether oligomers formed during heating of virgin olive oil show agglutination activity against human red blood cells. J Agric Food Chem 62(4):867–874. doi:10.1021/jf403745x

Pearce JM, Komoroski RA (1993) Resolution of phospholipid molecular species by 31P NMR. Magn Reson Med 29(6):724–731. doi:10.1002/mrm.1910290603

Pfeffer P, Luddy F, Unruh J, Shoolery J (1977) Analytical ^{13}C NMR: a rapid, nondestructive method for determining the cis,trans composition of catalytically treated unsaturated lipid mixtures. J Am Oil Chem Soc 54(9):380–386. doi:10.1007/bf02802040

Pignitter M, Somoza V (2012) Critical evaluation of methods for the measurement of oxidative rancidity in vegetable oils. J Food Drug Anal 20(4):772–777. doi:10.6227/jfda.2012200305

Pokorný J, Tài PT, Pařízková H, Šmidrkalová E, El-Tarras MFMM, Janíček G (1976) Lipid oxidation. Part 2. Oxidation products of olive oil methyl esters. Nahrung 20(2):141–148. doi:10.1002/food.19760200205

Reindl B, Stan HJ (1982) Determination of volatile aldehydes in meat as 2,4-dinitrophenylhydrazones using reversed-phase high-performance liquid chromatography. J Agric Food Chem 30(5):849–854. doi:10.1021/jf00113a014

Ribó JM, Crusats J, El-Hachemi Z, Feliz M, Sanchez-Bel P, Romojaro F (2004) High-resolution NMR of irradiated almonds. J Am Oil Chem Soc 81(11):1029–1033. doi:10.1007/s11746-004-1017-1

Ritota M, Marini F, Sequi P, Valentini M (2010) Metabolomic characterization of Italian sweet pepper (*Capsicum annum* L.) by means of HRMAS-NMR spectroscopy and multivariate analysis. J Agric Food Chem 58(17):9675–9684. doi:10.1021/jf1015957

Romano R, Giordano A, Paduano A, Sacchi R, Musso SS (2009) Evaluation of frying oil subjected to prolonged thermal treatment: volatile organic compounds (VOC) analysis by DHS-HRGC-MS and ^{1}H-NMR spectroscopy. Chem Eng Trans 17:879–884. doi:10.3303/cet0917146

Rossell JB (2001) Introduction. In: Rossell JB (ed) Frying. Woodhead Publishing, Sawston, pp 1–3. doi:10.1533/9781855736429.1

Sacchi R, Medina I, Aubourg SP, Giudicianni I, Paolillo L, Addeo F (1993) Quantitative high-resolution carbon-13 NMR analysis of lipids extracted from the white muscle of Atlantic tuna (*Thunnus alalunga*). J Agric Food Chem 41(8):1247–1253. doi:10.1021/jf00032a016

Saito H (1987) Estimation of the oxidative deterioration of fish oils by measurements of nuclear magnetic resonance. Agric Biol Chem 51(12):3433–3435. doi:10.1080/00021369.1987.10868583

Saito H, Nakamura K (1989) An improvement of the NMR method for estimating the oxidative deterioration of fish oils. Nippon Suisan Gakk 55(9):1663

Saito H, Nakamura K (1990) Application of the nmr method to evaluate the oxidative deterioration of crude and stored fish oils. Agric Biol Chem 54(2):533–534. doi:10.1080/00021369.1990.10869964

Saito H, Udagawa M (1992) Application of NMR to evaluate the oxidative deterioration of brown fish meal. J Sci Food Agr 58(1):135–137. doi:10.1002/jsfa.2740580122

Salih AM, Smith DM, Price JF, Dawson LE (1987) Modified extraction 2-thiobarbituric acid method for measuring lipid oxidation in poultry. Poult Sci 66(9):1483–1488. doi:10.3382/ps.0661483

Schaich KM (2005) Lipid oxidation: theoretical aspects. In: Bailey AE (ed) Bailey's industrial oil and fat products. John Wiley & Sons, Inc., Hoboken, NJ. doi:10.1002/047167849X.bio067

Schaich KM (2012) Thinking outside the classical chain reaction box of lipid oxidation. Lipid Technol 24(3):55–58. doi:10.1002/lite.201200170

Schaich KM (2013a) Chapter 1 - Challenges in elucidating lipid oxidation mechanisms: when, where, and how do products arise? In: Logan A, Nienaber U, Pan X (eds) Lipid oxidation. AOCS Press, Urbana, IL, pp 1–52. doi:10.1016/B978-0-9830791-6-3.50004-7

Schaich KM (2013b) Challenges in analyzing lipid oxidation: are one product and one sample concentration enough? In: Logan A, Nienaber U, Pan X (eds) Lipid oxidation. AOCS Press, Urbana, IL, pp 53–128. doi:10.1016/B978-0-9830791-6-3.50005-9

Schulte E (2004) Economical micromethod for determination of polar components in frying fats. Eur J Lipid Sci Technol 106(11):772–776. doi:10.1002/ejlt.200401004

Sebedio JL, Septier C, Grandgirard A (1986) Fractionation of commercial frying oil samples using sep-pak cartridges. J Am Oil Chem Soc 63(12):1541–1543. doi:10.1007/bf02553080

Senanayake SPJN, Shahidi F (1999) Oxidative deterioration of borage and evening primrose oils as assessed by NMR spectroscopy. J Food Lipids 6(3):195–203. doi:10.1111/j.1745-4522.1999.tb00143.x

Shahidi F, Wanasundara U, Brunet N (1994) Oxidative stability of oil from blubber of harp seal (*Phoca groenlandica*) as assessed by NMR and standard procedures. Food Res Int 27(6):555–562. doi:10.1016/0963-9969(94)90141-4

Shahidi F, Zhong Y (2005) Lipid oxidation: measurement methods. In: Bailey AE (ed) Bailey's industrial oil and fat products. John Wiley & Sons, Inc., Hoboken, NJ. doi:10.1002/047167849X.bio050

Shahidi F, Zhong Y (2010) Lipid oxidation and improving the oxidative stability. Chem Soc Rev 39(11):4067–4079. doi:10.1039/b922183m

Shantha NC, Decker EA (1994) Rapid, sensitive, iron-based spectrophotometric methods for determination of peroxide values of food lipids. J AOAC Int 77(2):421–424

Siddiqui AJ, Musharraf SG, Choudhary MI, Rahman AU (2017) Application of analytical methods in authentication and adulteration of honey. Food Chem 217:687–698. doi:10.1016/j.foodchem.2016.09.001

Simmler C, Napolitano JG, McAlpine JB, Chen SN, Pauli GF (2014) Universal quantitative NMR analysis of complex natural samples. Curr Opin Biotechnol 25:51–59. doi:10.1016/j.copbio.2013.08.004

Sinnhuber RO, Yu TC, Yu TC (1958) Characterization of the red pigment formed in the 2-thiobarbituric acid determination of oxidative rancidity. J Food Sci 23(6):626–634. doi:10.1111/j.1365-2621.1958.tb17614.x

Skiera C, Steliopoulos P, Kuballa T, Diehl B, Holzgrabe U (2014) Determination of free fatty acids in pharmaceutical lipids by ^{1}H NMR and comparison with the classical acid value. J Pharm Biomed Anal 93:43–50. doi:10.1016/j.jpba.2013.04.010

Skiera C, Steliopoulos P, Kuballa T, Holzgrabe U, Diehl B (2012a) ^{1}H-NMR Spectroscopy as a new tool in the assessment of the oxidative state in edible oils. J Am Oil Chem Soc 89(8):1383–1391. doi:10.1007/s11746-012-2051-9

Skiera C, Steliopoulos P, Kuballa T, Holzgrabe U, Diehl B (2012b) Determination of free fatty acids in edible oils by ^{1}H NMR spectroscopy. Lipid Technol 24(12):279–281. doi:10.1002/lite.201200241

Sochor J, Ruttkay-Nedecky B, Adam V, Hubalek J, Kizek R, Babula P (2012) Automation of methods for determination of lipid peroxidation. InTech, Rijeka. doi:10.5772/45945

Spyros A, Dais P (2000) Application of ^{31}P NMR spectroscopy in food analysis. 1. Quantitative determination of the mono- and diglyceride composition of olive oils. J Agric Food Chem 48(3):802–805. doi:10.1021/jf9910990

Spyros A, Philippidis A, Dais P (2004) Kinetics of diglyceride formation and isomerization in virgin olive oils by employing ^{31}P NMR spectroscopy. Formulation of a quantitative measure to assess olive oil storage history. J Agric Food Chem 52(2):157–164. doi:10.1021/jf030586j

St. Angelo AJ, Ory RL, Brown LE (1975) Comparison of methods for determining peroxidation in processed whole peanut products. J Am Oil Chem Soc 52(2):34–35. doi:10.1007/bf02901817

Summo C, Caponio F, Paradiso VM, Pasqualone A, Gomes T (2010) Vacuum-packed ripened sausages: evolution of oxidative and hydrolytic degradation of lipid fraction during long-term storage and influence on the sensory properties. Meat Sci 84(1):147–151. doi:10.1016/j.meatsci.2009.08.041

Sun X, Moreira RG (1996) Correlation between NMR proton relaxation time and free fatty acids and total polar materials of degraded soybean oils. J Food Process Preserv 20(2):157–167. doi:10.1111/j.1745-4549.1996.tb00852.x

Sun Y-E, Wang W-D, Chen H-W, Li C (2011) Autoxidation of unsaturated lipids in food emulsion. Crit Rev Food Sci Nutr 51(5):453–466. doi:10.1080/10408391003672086

Todt H, Guthausen G, Burk W, Schmalbein D, Kamlowski A (2006) Water/moisture and fat analysis by time-domain NMR. Food Chem 96(3):436–440. doi:10.1016/j.foodchem.2005.04.032

Tyl CE, Brecker L, Wagner K-H (2008) ^{1}H NMR spectroscopy as tool to follow changes in the fatty acids of fish oils. Eur J Lipid Sci Technol 110(2):141–148. doi:10.1002/ejlt.200700150

Valdés AF, Garcia AB (2006) A study of the evolution of the physicochemical and structural characteristics of olive and sunflower oils after heating at frying temperatures. Food Chem 98(2):214–219. doi:10.1016/j.foodchem.2005.05.061

Vlahov G (1996) Improved quantitative ^{13}C nuclear magnetic resonance criteria for determination of grades of virgin olive oils. The normal ranges for diglycerides in olive oil. J Am Oil Chem Soc 73(9):1201–1203. doi:10.1007/bf02523385

Wanasundara UN, Shahidi F (1993) Application of NMR spectroscopy to assess oxidative stability of canola and soybean oils. J Food Lipids 1(1):15–24. doi:10.1111/j.1745-4522.1993.tb00231.x

Wanasundara UN, Shahidi F, Jablonski CR (1995) Comparison of standard and NMR methodologies for assessment of oxidative stability of canola and soybean oils. Food Chem 52(3):249–253. doi:10.1016/0308-8146(95)92819-6

Wang X-Y, Yang D, Zhang H, Jia C-H, Shin J-A, Hong S, Lee Y-H, Jang Y-S, Lee K-T (2014) Antioxidant activity of soybean oil containing 4-vinylsyringol obtained from decarboxylated sinapic acid. J Am Oil Chem Soc 91(9):1543–1550. doi:10.1007/s11746-014-2492-4

Wheeler DH (1932) Peroxide formation as a measure of autoxidative deterioration. Oil Soap 9(4):89–97. doi:10.1007/bf02553782

White PJ (1995) Conjugated diene, anisidine value, and carbonyl values analyses. In: Warner K, Eskin NAM (eds) Methods to assess quality and stability of oils and fat-containing foods. AOCS Press, Champaign, IL, pp 159–178. doi:10.1201/9781439831984.ch9

Winkler-Moser J, Rennick K, Hwang H-S, Berhow M, Vaughn S (2013) Effect of tocopherols on the anti-polymerization activity of oryzanol and corn steryl ferulates in soybean oil. J Am Oil Chem Soc 90(9):1351–1358. doi:10.1007/s11746-013-2279-z

Winkler-Moser JK, Hwang H-S, Bakota EL, Palmquist DA (2015) Synthesis of steryl ferulates with various sterol structures and comparison of their antioxidant activity. Food Chem 169: 92–101. doi:10.1016/j.foodchem.2014.07.119

Zhang Q, Qin W, Li M, Shen Q, Saleh ASM (2015) Application of chromatographic techniques in the detection and identification of constituents formed during food frying: a review. Compr Rev Food Sci Food Saf 14(5):601–633. doi:10.1111/1541-4337.12147

Zimmerman DC (1966) A new product of linoleic acid oxidation by a flaxseed enzyme. Biochem Biophys Res Commun 23(4):398–402. doi:10.1016/0006-291X(66)90740-6

Printed in the United States
By Bookmasters